W9-ADY-843

WITHDRAWN
COLLEGE LIBRARY

PREBIOTIC AND BIOCHEMICAL EVOLUTION

Prebiotic and Biochemical Evolution

edited by

A. P. KIMBALL and J. ORÓ

*Department of Biophysical Sciences, University of Houston,
Houston, Texas, U.S.A.*

1971

NORTH-HOLLAND PUBLISHING COMPANY, AMSTERDAM-LONDON
AMERICAN ELSEVIER PUBLISHING COMPANY, INC. – NEW YORK

CARL A. RUDISILL LIBRARY
LENOIR RHYNE COLLEGE

© North-Holland Publishing Company – 1971

All rights reserved. No part of this publication may be reproduced, stored in a retrieval system, or transmitted, in any form or by any means, electronic, mechanical, photocopying, recording or otherwise, without the prior permission of the copyright owner.

Library of Congress Catalog Card Number 77-146184
North-Holland ISBN 0 7204 4090 4
American Elsevier ISBN 0 444 10074 1

Publishers:

North-Holland Publishing Company – Amsterdam
North-Holland Publishing Company, Ltd. – London

Sole distributors for the U.S.A. and Canada:

American Elsevier Publishing Company, Inc.
52 Vanderbilt Avenue, New York, N.Y. 10017

574.192
K56p
86458
Jan 1974

Printed in the Netherlands

Dedicated to the Memory of
Gerhard Felix Schramm
Born in Yokohama, Japan, June 27, 1910,
and Deceased in Tübingen, February 3, 1969.

PREFACE

More than one hundred years ago Ernst Haeckel proposed the nineteenth century version of the hypothesis of the "spontaneous" generation of life. At that time it was a logical extension of Wallace's and Darwin's concepts on the origin of species but had no support from either the physical, chemical or biological sciences. It is well known that the first comprehensive formulation of a theory of the origin of life in physical and chemical terms was presented by Oparin in 1924 and that the first experimental demonstration of the synthesis of amino acids under possible primitive earth conditions in support of this theory was published by Miller in 1953. Since then a substantial number of investigations have been carried out in this field of research, most of which have been collected in two volumes published in 1959 and 1965.*

The present volume is a collection of papers of research work performed during the last few years and assembled for the purpose of maintaining a continuity in this field of research. It begins by presenting important new developments in the area of abiotic synthesis of biochemical compounds and model systems of protocellular organization which attempt to trace different stages of chemical evolution before life appeared on our planet. The book then brings together some of the most relevant advances in the area of biochemical evolution which have been made by studying the nucleic acids, proteins, metabolic pathways, and other processes of living organisms. It ends, among other interesting papers on nucleic acids and enzymes, with the first successful attempt at the total synthesis of a naturally occurring polypeptide of 55 amino acids.

This chemical and biochemical research has progressed to a point where a meaningful and self-consistent picture of prebiotic and biochemical evolution starts to emerge from the experimental work done at different laboratories. We stand now on the brink of another phase of molecular evolution. We are rapidly accumulating knowledge and methods from which ways can be developed to predetermine our own evolutionary history.

This book developed from a Symposium on Proteins and Nucleic Acids (Synthesis, Structure, and Evolution) held at the University of Houston, Houston, Texas, in April, 1968, and from papers on Biochemical Evolution

---

* Oparin, A. I., Ed., The origin of life on the earth (Pergamon Press, Oxford, 1959).
Fox, S. W., Ed., The origins of prebiological systems and their molecular matrices (Academic Press, New York, 1965).

presented at the 6th Meeting of the Federation of European Biochemical Societies, held in Madrid, in April, 1969. We are indebted to the participants who have contributed to this volume and to several journals who have permitted us to reprint pertinent papers. Some of the authors of the papers delivered at the Houston meeting were kind enough to submit updated versions. The different contributions are not only relevant today but it is hoped they will be relevant many years from now. We wish to thank Drs. R. B. Hulbert, D. N. Ward, M. D. Anderson Hospital and Tumor Institute, Houston, Texas; Dr. S. Kit, Baylor University College of Medicine, Houston, Texas; and Dr. J. R. Cox, University of Houston, Houston, Texas; for their help with the Houston meeting; and Drs. A. Sols, C. Asensio, and Prof. A. I. Oparin, for their help with the Madrid meeting. We appreciate the assistance and encouragement received in the organization of the meetings and the preparation of the book from Drs. A. H. Bartel, A. Zlatkis, R. Segura, Mrs. Marianna O'Rourke, Thomas Spoor and Henry Simon, of the University of Houston. We also acknowledge the support and encouragement received from Dr. R. S. Young and the National Aeronautics and Space Administration.

A. P. KIMBALL

J. ORÓ

CONTENTS

M = Meeting of the Federation of European Biochemical Societies, Madrid,
   1969.
H = Symposium on Proteins and Nucleic Acids, Houston, 1968.

COACERVATE DROPS AS MODELS OF PREBIOLOGICAL SYSTEMS

A. I. OPARIN

Bakh Institute of Biochemistry, Academy of Sciences of the USSR, Moscow, USSR

Model experiments on the abiogenic synthesis of various organic substances (monomers and polymers) under the conditions similar to those which existed on the primitive Earth have led to a conjecture that long before the emergence of life on the Earth numerous biologically important compounds had been formed on our planet. This hypothesis was reliably confirmed by the discovery of similar abiogenically formed organic substances in meteorites and deep layers of the Earth's crust.

The solution of those substances in waters of the Earth's primitive hydrosphere (the so-called primordial broth) was the initial material for the development of life. However, the formation of primary organisms (eobionts) in the broth could proceed only via self-formation of relatively simple individual systems (protobionts) which in the course of long-term evolution came to be the precursors of all the living organisms on the Earth.

A good number of the systems capable of self-formation in the solution of organic substances similar to the primordial broth (bubbles of Goldacre, microspheres of Fox, coacervates of Bungenberg-de-Jong, etc.) can be not only conceptually developed but also practically reproduced in model experiments. Sometimes the self-formation is accompanied by the appearance of structures resembling those of living objects. But the important fact is the dynamic character of stability of these systems rather than this external resemblance.

The typical feature of living objects is that they are not static but stationary open systems in which continuous decay is constantly compensated for by substances and energy of the environment.

Initial protobionts should have had this stationarity. Their structure was significant for their further development only if it determined their dynamic stability and in some cases if it also dictated the growth of a given open system at the expense of its interaction with the environmental medium.

From this viewpoint, coacervate drops seem to be the most promising (though not the only possible) models for studying evolution of protobionts. In the medium similar to the primordial broth coacervates are readily self-formed via aggregation of varied polymers (e.g., polypeptides and polynucleotides) to drops of microscopic size. Of key importance for the self-formation is the level of polymerization but not the intramolecular structure of polymers. Fig. 1 presents, for instance, drops formed from monotonously built polymers (polyadenine and polylysine).

The drops are separated from the surrounding solution by a well outlined surface but are capable of interacting with the surrounding medium, selectively adsorbing such substances as amino acids, sugars, mononucleotides and releasing products of reactions developing within them.

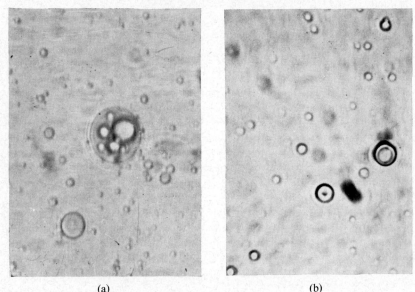

(a)   (b)
Fig. 1.   Coacervate drops containing polypeptides and polynucleotides.
(a) RNA + polylysine;   (b) Poly-A + polylysine.

Without this interaction coacervate systems being static systems, are unstable. For instance, polymers are readily decomposed both spontaneously and under the influence of catalysts included in the drops. As a result, the drops disintegrate partially, as can be seen in the electron-micrograph, or disappear completely (fig. 2).

Despite disintegration processes, coacervate drops can long persist, acquiring dynamic stability due to the entry of matter and energy from the environment. In the simplest event, the disintegration occurring in the drop can be compensated for by ready-made polymers entering the system from the surrounding medium [in our experiments at the expense of a polymer (RNA) added to the equilibrium fluid].

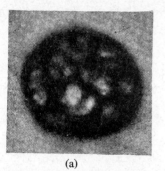
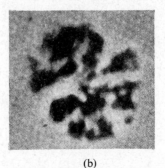

(a)   (b)

Fig. 2.   The coacervate drops containing serum albumin, gum arabic, RNA and RNA-ase.
(a) no incubation;   (b) 20 min incubation.

In other cases, drops adsorb energy-rich monomers, and their dynamic stability depends on the polymer synthesis occurring within the system. If the synthesis rate is balanced up by the disintegration rate, drops develop dynamic stability and can persist for a long time; if synthesis prevails over disintegration, drops increase their volume and weight, growing before our eyes.

Here is a scheme of one of the experiments (fig. 3).

A coacervate drop, formed from histone and gum arabic, is placed into the glucose-1-phosphate solution. Under the influence of glucosyl-transferase incorporated in the drop, glucose-1-phosphate transforms into starch, at the expense of which the drop grows, if the synthesis rate (a) exceeds the disintegration rate (b). This scheme illustrates another procedure.

In fig. 4 a drop consisting of histone and RNA adsorbs ADP from the surrounding medium. Due to the polymerization reaction (poly-A formation) the drop growth occurs which can be easily recorded.

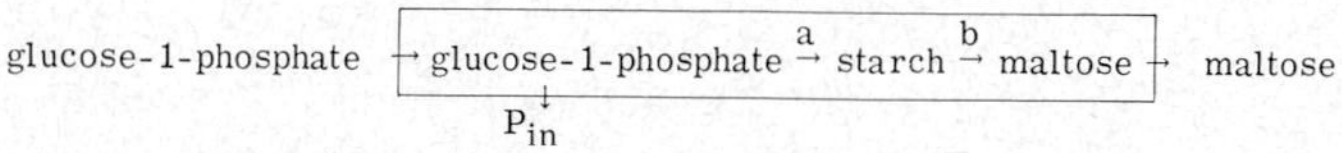

$$\text{glucose-1-phosphate} \dashv \underset{P_{in}^+}{\underbrace{\text{glucose-1-phosphate} \overset{a}{\to} \text{starch} \overset{b}{\to} \text{maltose}}} \vdash \text{maltose}$$

Fig. 3.   Synthesis and hydrolysis of starch in the coacervate drop.

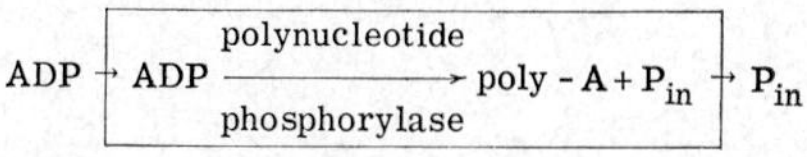

$$\text{ADP} \dashv \underset{\text{phosphorylase}}{\overset{\text{polynucleotide}}{\boxed{\text{ADP} \longrightarrow \text{poly-A} + P_{in}}}} \vdash P_{in}$$

Fig. 4.   Scheme of polyadenine synthesis by polynucleotide phosphorylase in the coacervate drop.

In the examples described the rate of reactions developing in the drop is mainly dictated by the activity of incorporated catalysts (in laboratory experiments: enzymes). However, the indicated rate can depend on other events, e.g., on the light energy influx from the environment, occurrence of photosensitizers in drops and, finally, their intermolecular organization. This can be exemplified by protein-lipid coacervates incorporating sensitizers of reduction–oxidation photoreactions. Fig. 5 shows coacervate drops consisting of protein and potassium oleate with chlorophyll incorporated.

Under the influence of light the photoreaction of ascorbic acid oxidation and methyl red reduction develops in the drop according to the following scheme: fig. 6.

In this system the photoreaction rate in the drop is 60 times as high as in the equilibrium fluid. This occurs mainly due to an increase of the concentra-

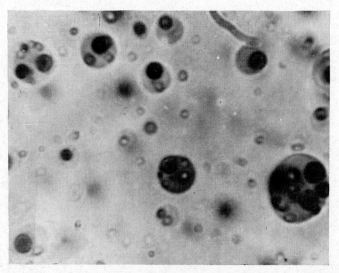

Fig. 5.   The coacervate drops containing serum albumin and oleate potassium, including chlorophyll.

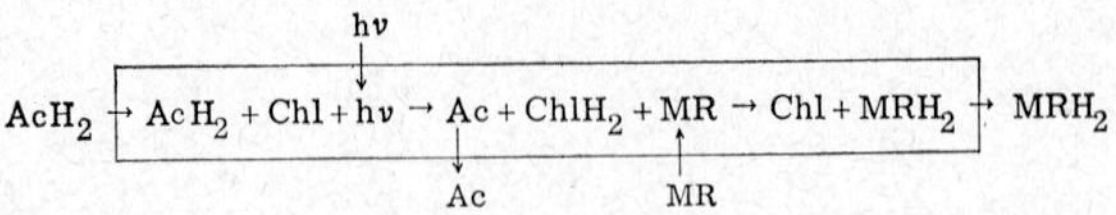

$$AcH_2 \rightarrow AcH_2 + Chl + h\nu \rightarrow Ac + ChlH_2 + MR \rightarrow Chl + MRH_2 \rightarrow MRH_2$$

Fig. 6.   Scheme of ascorbic acid oxidation in the coacervate drop, containing chlorophyll (red light).

tion of the sensitizer in coacervate drops which can be demonstrated by direct assay.

The situation is different if coacervate drops involve lecithin. In this case the pigment location on phosphatic micellae becomes orderly as a result of a fixed position of the phytol tail of chlorophyll between two hydrophobic residues of lecithin. This more perfect spatial intermolecular organization of coacervate drops increases immediately the rate of the reactions by two orders of magnitude. In this event the photoreaction proceeds approximately by 100 times faster than it is required for the pigment concentration. In the above case the role of the phytol group is indicated by the fact that the chlorophyll replacement by the phytol-free pigment – porphyrin IX – does not lead to the indicated increase of the photoreaction rate in the lecithin coacervate.

The intermolecular spatial organization seems to be an intermediate stage in the course of development of structures visible under the electronic microscope.

The effect of the spatial organization on the rate of reactions occurring in the coacervate drop can be also distinctly seen if two or more reactions proceed in the system. As an example we reproduced the beginning of the pentose cycle in coacervate drops composed of lecithin and enzyme proteins in which the hexokinase and glucose-6-phosphate dehydrogenase action was combined (fig. 7).

The diagram in fig. 8 presents curves characterizing the development of the reactions in homogeneous and heterogeneous conditions.

We can see that, if the two reactions develop together under homogeneous conditions, then the rate of the first reaction increases due to the presence of glucose-6-phosphate dehydrogenase which oxidizes glucose-6-phosphate formed in the first reaction. The acceleration is, however, comparatively low. Simple incorporation of hexokinase into coacervate drops enhances significantly the effect of the enzyme as compared to that of homogeneous conditions. This reveals the role of the spatial organization. It becomes apparent upon a combination of the effect of the two enzymes in coacervate drops. In such a case the combined effect greatly exceeds that observed in homogeneous

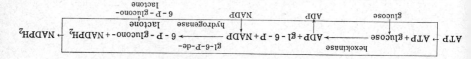

Fig. 7. First steps of pentosephosphate cycle in the coacervate drop.

conditions. A similar phenomenon was detected in the case of a combined effect of hexokinase and polynucleotide phosphorylase, as shown in the scheme (fig. 9).

Here we can clearly see the effect of the spatial organization on the rate of the polymer synthesis, and therefore on the rate of the growth of coacervate drops.

On the basis of the above scheme, we carried out an experiment with two differently organized systems which were built of histone and enzyme proteins. They differed in the fact that one of them incorporated glucose and a

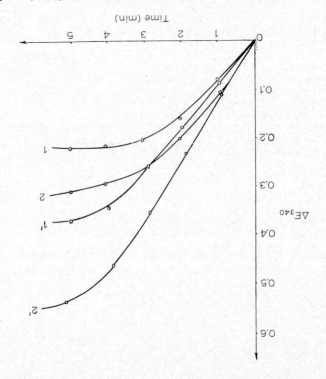

Fig. 8.   The first step of pentose cycle in the coacervate drops (curve 2').

Fig. 9.   Scheme of polyadenine synthesis by polynucleotide phosphorylase in the coacervate drop in presence of hexokinase.

combination of two enzymes (polynucleotide phosphorylase and hexokinase) whereas the other involved polynucleotide phosphorylase alone. Both drops were placed into a solution of commercial ATP which contained up to 20% of ADP (fig. 10).

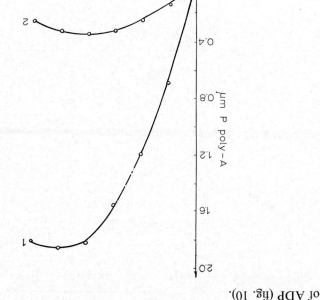

Fig. 10. The dependence of the synthesis of poly-A by polynucleotide phosphorylase in the coacervate drops on the time of incubation. (1) histone/polynucleotide phosphorylase/hexokinase; (2) histone/polynucleotide phosphorylase.

This diagram shows that drops having more perfect internal organization (curve 1) grow much faster than those having less perfect organization (curve 2) though they are in the same environment.

Here we can already distinguish the beginnings of new principles of development originating at the junction of chemical and biological evolution, the beginnings of prebiological natural selection. These principles underlie the entire further evolution of protobionts.

SELF-ASSEMBLY OF THE PROTOCELL FROM A
SELF-ORDERED POLYMER*

SIDNEY W. FOX

Institute of Molecular Evolution,
University of Miami, Coral Gables, Florida, U.S.A.

The problem of the origin of life, or in truly perceptive nineteenth century terms, the problem of spontaneous generation, has often been regarded as one of overwhelming complexity. Upon analysis, with the aid of hindsight, this problem loses some of its imponderability. The aspect of evolution which first received major attention was that of the progression, in principle, from primitive cell to contemporary cell and to contemporary multicellular organisms. This stage is the one that has been illuminated mechanistically by Darwin's theory of selection. We can now regard this stage as far more intricate and involved than the emergence of primitive life from the primordial reactant gases. By such an analysis, the primordial cell is emphasized, the highly ramified later stages are removed from purview, and the limits of the meaningful problem are identified.

The preorganismic stage can also be analyzed. For intellectual convenience, it may be divided into two or three parts. The first of these parts is that of the spontaneous organic synthesis involved in the production of the small organic molecules which are necessary for contemporary and, presumably for, primitive organisms. The second step is the spontaneous synthesis of the polymers and of cells. This latter constitutes in turn, however, two stages. These two steps were collectively most forbidding in quality, and are particularly significant to life and therefore, to its origin. Our most modern knowledge requires that we recognize that a primitive cell can not be a synthesized entity in the true meaning of "synthesized". The precursor macromolecule can be conceived of as synthetic. When the appropriate macromolecule has been formed, the final and crucial stage, leading to a primitive organism, would then be one of self-assembly. The term "self-assembly" and the con-

* Since 1960, this research has been aided by the National Aeronautics and Space Administration, currently Grant no. NsG-689. Contribution no. 096 of the Institute of Molecular Evolution.
Paper presented to International Convention of Biochemists, Bangalore, India, 7 September 1967. (Reprinted by permission from the Journal of Scientific and Industrial Research (India).)

cept have recently been receiving increasing recognition e.g. [1] in the bio-chemistry of contemporary systems.

One way in which students of the total problem have dealt with the seemingly great complexity has been to postulate a long chemical evolution [2] extending over, say, 25 million years. I will explain here why our expe-riments lead to the interpretation that the essential steps from primordial gases → amino acids → primitive protein → a primitive organized structure having simultaneously many lifelike properties including the ability to parti-cipate in its reproduction, could have occurred many times in a very short period, say 25 hr. My immediate problem is to present the salient experimen-tal material in an even shorter time. This problem exists because of the careful devotion to it by many associates during 14 years of continuous study in our laboratory. Accordingly, I shall rely heavily on summaries and upon exam-ples from our laboratory and others.

Our approach to this problem was based on clues from contemporary cells. The results of experiments have been evaluated in part by how well they lead to an increasing appearance of the properties that are associated with contem-porary cells. The experiments have however been based on very simply derived initial systems and simple processes. These employ conditions that have proved to be plausible not only for geologically ancient times; the con-ditions identified are widespread now through recorded history [3].

Models of the prebiotic synthesis of small organic compounds such as monosaccharides, amino acids, purine and pyrimidine bases, ATP, porphy-rins, etc. have been described from many laboratories including those of Ponnamperuma [4]. Oró [5], Orgel [6] and our own [7]. Since the essence of life is generally recognized as being that of the biopolymers, protein and nucleic acid, this paper will focus on questions involving the primordial formation of protein and nucleic acid and on the attributes of the polymers formed in the laboratory. It will deal also with complexes of the two.

Turning first to the question of proteins, we find that a number of studies of the synthesis of peptide bonds, mostly in aqueous solution, have been carried out in a number of laboratories. Akabori [8] employed the pro-gressive substitution of polyglycine as a model of the first protein, and Matthews [9] has reported a similar process. The model of our laboratory which relies on heat and hypohydrous conditions is the only one that has yielded polymers of molecular weight in the thousands, a content of all eighteen amino acids common to protein, several protoenzymic activities, and it is the only model which has been demonstrated to yield organized structures with a lengthy roster of the properties of the contemporary cell [10].

This synthesis has the simplicity appropriate to geologically spontaneous occurrences, and it yields both polymer and organized units in abundance.

One other synthesis, which has most of the attributes enumerated, uses as intermediates the reactive Leuchs anhydrides of the amino acids. This synthesis was also first performed in our laboratory, by Hayakawa [11]. Of these two syntheses, only the thermal process has the simplicity appropriate to geologically spontaneous occurrences.

The thermal syntheses, first attempted in 1953, were indicated as thermodynamically possible following studies of Borsook, Huffman, Ellis and Fox [12]. The results of calculations from the tabulated physical constants have shown that one could expect in an open aqueous solution only small yields of small peptides unless the reaction were somehow coupled to an endergonic one.

The fact that organisms are nearly always aqueous entities has led some to assume that hot, dry conditions would not have been appropriate to early life and they have somehow projected such thinking to precursor molecules. Our chemical experience however tells us that macromolecules can easily survive conditions lethal for ordinary cells, and our biological experience reminds us that bacterial spores are relatively resistant to heat and dryness.

The reaction involving formation of peptide bond with its attendant Gibbs free energy change is:

$$H_2NCHRCOOH + H_2NCHR'COOH$$
$$= H_2NCHRCONHCHR'COOH + H_2O$$
$$\Delta G^0 = 2000 \text{ to } 4000 \text{ cal.}$$

As the number of peptide bonds per molecule increases, the equilibrium constant becomes geometrically more unfavorable. Dixon and Webb [13] have calculated that the volume of 1 M amino acid solution in equilibrium with one molecule of protein of molecular weight 12,000 would be 10^{50} times that of the Earth! Stated otherwise, uncoupled synthesis from amino acids in water should be expected to give small yields of small peptides only.

In order to shift this equilibrium to favor synthesis, one can postulate removal of either product. Theoretically, one contribution of a membrane in contemporary protein-synthesizing systems may be the overcoming of an energy barrier by separation of synthesized macromolecules from the aqueous solution. This process could not apply, however, until a membrane composed of macromolecules had first formed. Our attention therefore shifts to removal of the other product, water. This route to peptide bond synthesis can be visualized as a geochemical possibility. It has also been demonstrated experimentally [7].

One mode of removal of water, as thus suggested by the thermodynamic analysis, would be that of heating the amino acids above the boiling point of water. When this possibility was initially contemplated, the probability of gross decomposition had to be considered. Such a consequence of heating α-amino acids above the boiling point of water has been recorded in the literature a number of times and was also common knowledge. We were led to attempt the thermal condensation by employing an inference from comparative studies of organismic protein, the fact that the amino acids which most dominate the composition of proteins are glutamic acid and aspartic acid [7]. These contents were taken hypothetically to be an evolutionary reflection of a circumstance required for the primordial formation of prebiotic protein.

Another consideration that had to be dealt with was the somewhat vague feeling that, without nucleic acids present, the necessary systematic sequences of amino acid residues would not result. This problem was conceptually eliminated, in principle, by studies of enzymic acylpeptideanilide synthesis [14] which demonstrated that interactions of amino acid residues would alone select the sequence formed. (This principle and the inference that prior nucleic acids may have been unnecessary [7] has since been corroborated by Steinman [15] in another system of reacting amino acids.) Since the difficulties were thus conceptually surmountable in 1953, heating was employed (fig. 1). The discussion now deals with experimental observations.

A typical thermal condensation used at first a mixture of 1 part of aspartic acid, 1 of glutamic acid, and 1 of an equimolar mixture of the 16 other amino acids common to protein. This mixture was heated at 170°C for 6 hrs [16]. The resulting light amber glassy product, not depicted, is entirely soluble in water by salting-in, and can be purified by salting-out. Such products yield amino acids 100% by acid hydrolysis, they contain some proportion of each of the amino acids common to protein (or fewer as desired), and molecular weights of many thousands. They have many other properties of protein and are called proteinoids. The proteinoid described in this example is, because of the proportions reacted, a 1:1:1 type. More recently, Waehneldt has shown in our laboratory that aspartic acid and glutamic acid may be merely equimolar with the 16 other amino acids. Proteinoids are produced even so. Yields in the usual syntheses are typically in the range of 10–40%, higher yields being obtained by the addition of phosphates [17, 18]. Many other laboratories have repeated this synthesis and have confirmed it and its simplicity, which is in turn crucial to the geological validity. The spontaneous occurrence of a carbobenzoxy synthesis or the

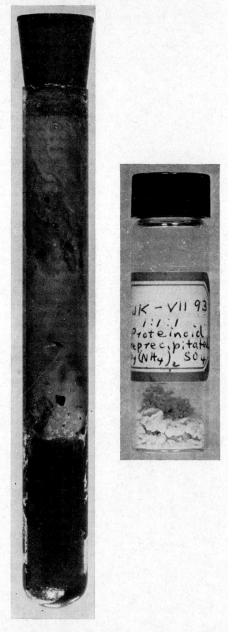

Fig. 1.   On left, tube containing mixture of amino acids heated to above the boiling point
of water. On right, granular polymer prepared by heating a mixture of amino acids
containing sufficient proportions of aspartic acid and glutamic acid.

formation and condensation of N-carboxy amino acid anhydrides cannot, of course, be defensibly imputed to the geological environment.

The numerous properties which the proteinoids have in common with proteins are described in detail in the literature, and these have been reviewed a number of times [7,10].

Table 1 lists the many simultaneous structural, chemical, and biological properties all of which are described in literature cited bibliographically, except for recently demonstrated hormonal activity and a few other aspects.

TABLE 1

Properties of thermally prepared proteinoids

Limited heterogeneity
Qualitative composition identical to that of protein
Quantitative compositions resembling those of proteins
Quantitative recoverability of amino acids upon hydrolysis
Range of molecular weights like those of smaller protein molecules
Positive color tests as for protein
Solubilities resembling those of protein classes (albumins, globulins, etc.)
Some optical activity
Tendency to be salted-in
Tendency to be salted-out
Precipitability by protein group reagents (phosphotungstic acid, etc.)
Hypochromicity
Infrared absorption maxima as found in protein
Some susceptibility to proteases
Nonrandom distribution of amino acid residues
Many catalytic activities
Inactivatability of catalytic power
Nutritive quality
Melanophore stimulating activity
Morphogenicity

Time will be devoted here to reviewing only three salient properties on which many new data are available: (1) limited heterogeneity and the related question of systematic sequences, (2) catalytic activity, and (3) the property of forming regular and highly structured particles.

The first indication that a thermal condensate of eighteen amino acids yields no more than two electrophoretic individuals was from a study of Vestling [16]. Subsequently Harada in our laboratory showed that two fractionations of a 2:2:3-proteinoid from hot water resulted in virtually no change in amino acid composition [19]. Also, N-terminal [16] or C-terminal analyses [20] of many polymers demonstrated marked disparities from the total compositions, indicating that systematic nonrandom sequentialization occurred. This could be due only to selective interactions of amino acids during thermal condensation. Since that finding, many new data obtained by Nakashima and a review of all of the data have been published [21].

The elution pattern from fractionation on DEAE-cellulose is shown in fig. 2.

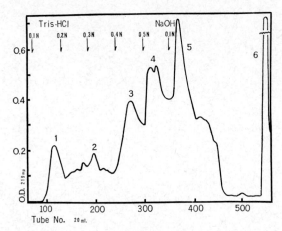

Fig. 2.   Distribution of 1:1:1-proteinoid on elution from a DEAE-cellulose column by tris-HCl buffer.

While no random assortment of polyanhydro-α-amino acids has been prepared such that a comparison would be possible, we would except theoretically an elevated nearly horizontal line for an elution pattern of a disordered polymer. What is repeatedly found, instead, is a pattern of six major peaks, of which some are already symmetrical.

For comparison is presented an elution pattern of turtle serum protein [22] also fractioned on DEAE-cellulose. Eight major peaks, with less spread in each peak, are observed. The individual peaks of the eluate sediment in the Spinco Model E (fig. 3) to indicate for the fractions the degree of ultracentrifugal homogeneity which is observed.

Fig. 4 shows the complete and partial hydrolyzates of three fractions from DEAE-cellulose. The bottom three are partial hydrolyzates in each group. The patterns are highly similar, as can be seen.

A total picture of nonrandom sequences in the linear distribution, discrete macromolecular fractions, and relatively uniform composition and sequences throughout the entire polymer is supported also by high voltage electro-

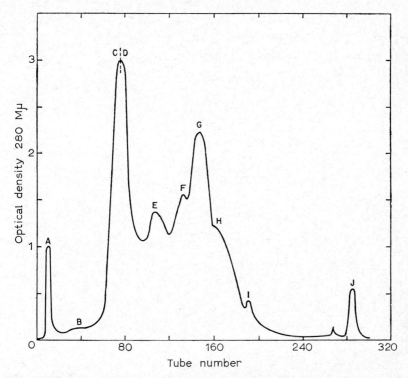

Fig. 3.   Distribution of turtle serum proteins on elution from a DEAE-cellulose column by sodium phosphate buffer.

phoresis, by separation on columns of Sephadex, by polyacrylamide fractionations, and by gel electrophoresis.

Reports of catalytic activity in proteinoids are summarized in table 2. These findings are from six laboratories and they include catalysis of the hydrolysis of esters and of ATP and decarboxylations of a number of natural substrates [23]. Krampitz has recently recorded an example of amination. Michaelis-Menten kinetics have been reported in several of the studies. The activities are mostly weak. As Calvin pointed out for the iron-containing enzymes [24] and as I pointed out in 1953 [25], weak primitive enzymic activity would be selected and enriched by Darwinian processes in organisms. Of relevance, also, is the fact that some gross specificities have been identified and that individual proteinoid preparations each have an array of catalytic activities. A metabolic sequence has been recorded, for example, for oxaloacetic acid → pyruvic acid → acetic acid + CO_2.

We can explain the evolutionary potential of polyanhydroamino acid

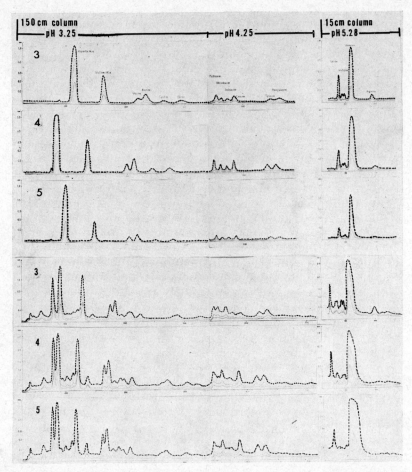

Fig. 4. Top 3 profiles of hydrolyzates of 3 fractions of proteinoidamide show great similarity in amino acid content. Bottom 3 profiles show similarity in peptides obtained on partial hydrolysis.

catalysts by the fact that in the same macromolecule are found not only a variety of chemically functional groups, but also the products of interaction of the fields of force of two or more of these groups. This picture of chemical polyfunctionality is also of course applicable to proteins and emphasizes what Needham [38] has referred to as "the uniqueness of biological materials". Proteinoid is in one sense perhaps even more unique than any one protein in that it is less specialized, and recalls thereby for serious consideration the concept of "urprotein" [38].

TABLE 2

Catalytic activities in proteinoids

Reaction	Salient Finding	References
p-Nitrophenyl acetate	$Activity_{ptd} > Activity_{hsd}$	[26]
p-Nitrophenyl acetate	Activities	[27]
p-Nitrophenyl acetate	Inhibition by organic phosphates	[28]
p-Nitrophenyl acetate	Detailed treatment	[29]
p-Nitrophenyl acetate	"Active site", and inactivation	[30]
Glucose → glucuronic acid → CO_2	First natural substrate reported	[31]
ATP → ADP, et al.	Biochemical energy source	[32, 33]
p-Nitrophenyl phosphate	Second phosphate hydrolysis	[34]
Pyruvic acid → acetic acid + CO_2	Decarboxylation M-M kinetics	[35, 33]
Oxaloacetic acid → pyruvic acid + CO_2	Catalyzed by ptds of type not active on pyruvic acid	[36]
α-Ketoglutaric acid + urea → glutamic acid	Proteinoid and Cu each needed	[37]

The property of forming structurally organized units on contact with water is crucial to a comprehensive theory of the origin of the first cell. This tendency is intrinsic to many thermal polymers of amino acids, as we reported in 1960. The need for a macromolecular precursor of the first cell has been emphasized in theoretical discussions by Oparin [40], by Wald [41], by Lederberg [42], and by others. The degree to which thermal proteinoid meets this need by providing organized microscopic units having numerous associated properties such as are found in contemporary protein and in contemporary cells could not have been predicted. Only some of the more salient properties will be presented here. Others are documented with references to supporting literature [43]. The properties include, among others, a cellular type of ultrastructure, double layers, abilities to metabolize, to grow in size, to proliferate, to undergo selection, to bind polynucleotides, and to retain some macromolecules selectively. The structures of fig. 5 are usually produced merely by heating the proteinoid with water or aqueous solution. The clear liquid, on cooling, deposits huge numbers of microscopic units, 0.5–80 μm in diameter, of quite uniform size in any one preparation. These are usually found as spherules, as in this photomicrograph, but they occur also as filaments, budded microspheres, as twinned units, and in other shapes. They are exceedingly numerous; one gram of heated amino acids can produce many billion spherules. As pointed out by us and by others [44], they have physical stability comparable to that of contemporary cells; they can, for instance, be sectioned for electron microscopy. In their uniform size and in other respects they are readily distinguished from oily droplets or from the usual

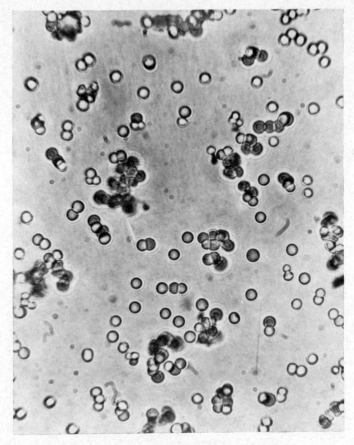

Fig. 5. Proteinoid microspheres.

coacervate droplets. While they can be produced as entirely separate units, they also tend to associate [45]. By adjustment of the basic amino acid content, they can be produced to stain either Gram-negative or Gram-positive [46]. They also have been shown to have some of the catalytic activities which have been carefully identified in the polymer of which they are composed [23].

The ultrastructure of the microsphere is shown in the electron micrograph of fig. 6. Above is a section of *Bacillus cereus* which has been fixed by osmium tetroxide, sectioned, and electron micrographed, by Murray [47]. In the lower figure is seen a section of a proteinoid microsphere fixed with osmium tetroxide and sectioned. While some bacteria reveal a more organized pattern

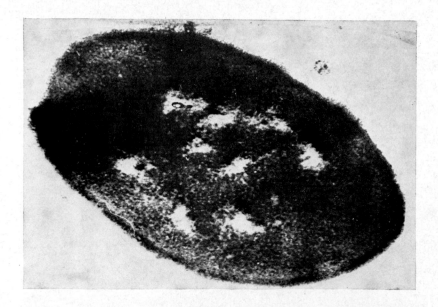

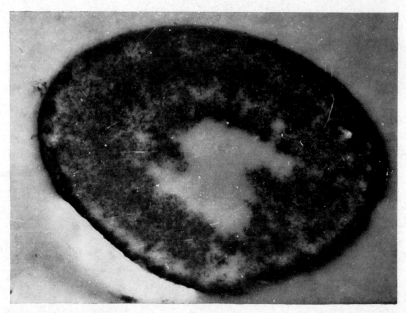

Fig. 6.   Electron micrographs of *Bacillus cereus* (above) and of proteinoid microsphere (below). Each has been fixed with osmium tetroxide and sectioned.

than *Bacillus cereus*, this micrograph reveals only a boundary and granular cytoplasm. The granular appearance is found also in the proteinoid micro-sphere, and the latter has a more definite boundary. In one place the boundary appears to be a double layer. In the experiment preceding the electron micrography of fig. 7, the polymer in the interior was first caused to diffuse out through the boundary by raising the pH, in the suspension of proteinoid

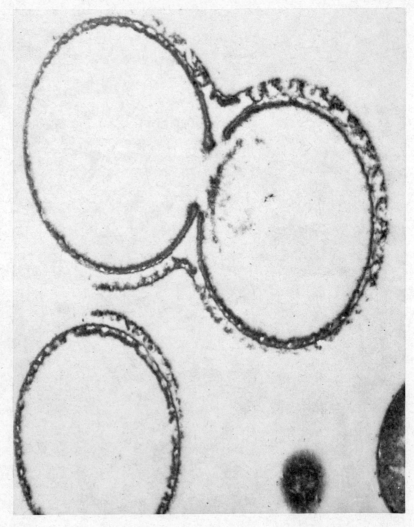

Fig. 7.   Proteinoid microsphere subjected to raised pH. Double layers are evident.

microspheres by one to two units. Double layers are clearly evident [7]. We can also see part of the results of other phenomena. The diffusion depicted is one manifestation of selective behavior in the boundary or membrane. This diffusion is further illustrated in fig. 8, in which the effect was followed by photographing in ultraviolet light through the quartz optics of Dr. Philip Montgomery's microscope at the University of Texas.

The proteinoid, like protein [48], absorbs at ultraviolet wavelengths. These pictures and related data indicate that polymer is not condensing on the membrane but is passing selectively through a membrane composed of very similar polymer, as has been shown by analysis [49].

Models of primitive polynucleotides have also been examined. With Waehneldt we have reported how nucleoside mono- and tri-phosphates could have been prebiotically synthesized in quantity [50], following an earlier report on the production of ATP by Ponnamperuma [51]. Thermal polymers of mononucleotides have been shown, by Schwartz, in our laboratory, to be attacked by ribonuclease and by venom phosphodiesterase [52]. Recently, we reported with Waehneldt and others that such polynucleotides, as well as calf thymus DNA and yeast RNA, bind with appropriate proteinoids to yield models of the various nucleoprotein particulates found in the contemporary cell, such as ribosomes, chromosomes, chromatin, etc. (fig. 9).

In fig. 10 is provided an example which shows fibers produced from lysine-rich proteinoid and calf thymus DNA. Very small microspheres result when RNA or thermal polyribonucleotides are used instead of DNA. The ratio of polynucleotide to basic proteinoid in such complexes tends to be quite constant. The fibrous and globular morphologies are reminiscent of contemporary analogs as in chromosomes and ribosomes respectively. Those proteinoids that bind to form such particles have a ratio of basic amino acid to dicarboxylic amino acid above a minimum of 1.0.

With polyphosphoric acid and temperatures of 60–100°C, experiments in our laboratory have shown that either mononucleotides or amino acids might be polymerized [18, 52]. Accordingly, these processes might ordinarily have occurred simultaneously. The suggestion of Calvin that proteins and nucleic acids might have arisen simultaneously in a primordial event [53] is thus consistent with the experimental demonstrations. Although model studies of prebiotic polynucleotides have been pursued, as indicated, a basic question that persists is that of how many properties models of primitive protein systems might display *without* polynucleotides. This question especially deserves to be asked in view of the fact that proteinoids contain their own information and have sharply limited variation without any control by

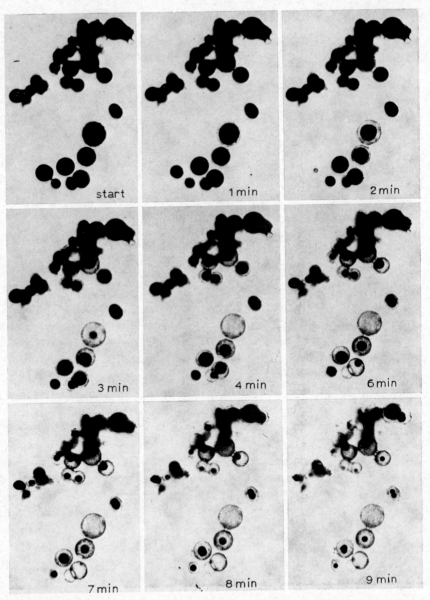

Fig. 8. *p*H effect followed by photograph in ultraviolet light through quartz optics.

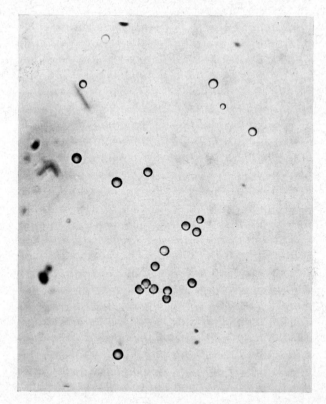

Fig. 9.   Microspheres composed of RNA and lysine-rich proteinoid.

nucleic acids. Also, the self-assembling properties of proteinoid yield ultra-structure, double layers, fission, etc., without nucleic acid control [7]. Can then, for example, proteinoid microspheres multiply without polynucleotides present? In figs. 11–14 we observe how in a very simple manner proteinoid microspheres do, in fact, participate in the reproduction of their own like-ness [43].

In the first photomicrograph are seen a number of proteinoid microspheres which have been allowed to stand in their mother liquor for two weeks. On these are found "buds" which in appearance resemble buds on yeast. We first observed such buds in 1959. These "buds" can grow in size either while attached to the parent microsphere, or after separation. Removal can be accomplished at various stages of growth in size by shock – electrically, thermally, or mechanically. In the first two modes, we believe that some

Fig. 10.   Photograph of fibrous complex of calf thymus DNA and of
lysine-rich proteinoid.

interfacial material is dissolved. "Buds" are seen in the second picture. In
order to demonstrate the next event rigorously, the separated buds were
stained with crystal violet. When the stained separated buds are allowed to
stand in a solution of proteinoid saturated at 37°C, and this is allowed to
cool to 25°C, the buds, in an appropriately sized vessel, "grow" by accretion
to the size shown within one hour. The resultant units tend to be very uniform

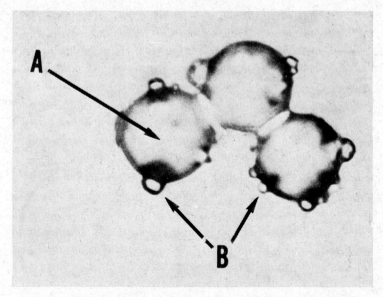

Fig. 11.   "Budded" proteinoid microsphere A: Microsphere B: "Bud".

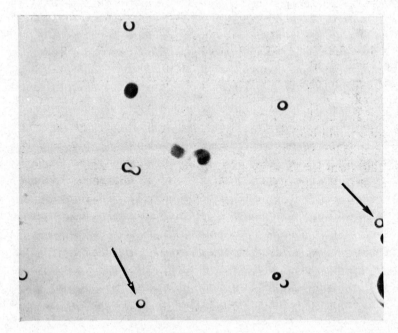

Fig. 12.   Separated "buds".

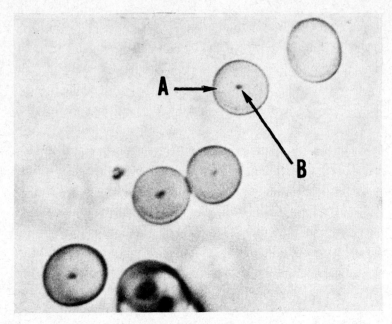

Fig. 13.   Accretion of proteinoid particle around crystal violet-stained "buds".
A: Proteinoid particle   B: Stained center.

in size; the opposing forces are evidently precisely balanced under any given set of conditions.

In this manner we can visualize how proteinoid microspheres could simply first have developed the ability to participate cyclically in the reproduction of their own likeness. In manifesting such a process, a primitive organized structure would be functioning as a nearly complete heterotroph in that it obtained its large molecules by feeding on the environment instead of synthesizing them itself. Many theorists on the subject of abiogenesis have reasoned that the first organisms must have been heterotrophs. One may find the arguments in the writings of Oparin [40], Haldane [54], Pirie [55], Pringle [56], Horowitz [57], Van Niel [58], and others. To repeat, the hypothetical need for nucleic acid-mediated constraints on primitive protein are seen not to apply because of the internal constraints on the primary structure of proteinoid [21].

Experiments in our laboratory have indicated how multiplication could occur also through the model of a primitive kind of binary fission [59] and growth by accretion, as well as through budding.

With cyclical proliferation due to budding or fission, as depicted, Darwi-

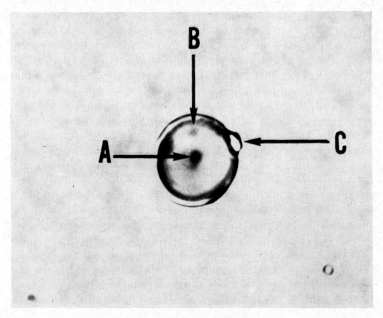

Fig. 14. "Daughter bud" on microsphere. A: Stained center. B: Microsphere by accretion. C: "Daughter bud".

nian selection can, conceptually, occur. Increasingly incisive experiments on Darwinian selection from nonbiological precursors have in fact been performed with acid proteinoid and neutral proteinoid.

The condensed description of this paper (in conjunction with other reports [7]) explains how a primitive organism capable of a kind of self-multiplication and possessing other salient properties could have emerged from primitive gases through the amino acids and subsequently through protein-like polymer. As perhaps need not be re-emphasized, this model of a kind of primitive unit is clearly not a contemporary cell, at least not of the usual contemporary organism.

From this model, however, we visualize that in evolving to a contemporary organism a primitive self-replicating heterotroph would especially have had to develop an internal synthesis of protein and of polynucleotide.

The model processes which have been described are extremely simple. They consist of (a) heating above the boiling point of water, and (b) the intrusion of water. This simple sequence requires (a) geologically anhydrizing temperatures, e.g. those above 100°C and (b) sporadic rain or other common geological events of water such as drought or recession of the seas. These conditions (a) and (b) have been widespread geologically and, in fact, are quite widespread on the Earth today. The reactions are rugged, their occurrence is not easily disturbed by added substances since they are not solutes in aqueous solution, and the products have long-term stability. All details in the processes are found to be sequentially compatible.

To summarize, I use a 1966 quotation from Lederberg [42], "The point of faith is: make the polypeptide sequences at the right time in the right amounts and the organization will take care of itself. This is not far from suggesting that a cell will crystallize itself out of the soup when the right components are present." The results described here (and others) show that when amino acids are simply and suitably heated, polypeptide sequences to at least some degree make themselves, and they do this in "the right amounts". The organization of the polypeptides does indeed take care of itself when water is added to thermal proteinoid, which appears to be the right component to crystallize out as a cell. Individual properties of proteins, polynucleotides, and of cells can be mimicked by other substances and units. Thermal proteinoid and its organized particles are the only chemically synthetic products, however, which have been shown to possess in *simultaneous* association and in life-like inclusiveness properties of the contemporary cell and its structural polymer, with indications of the potential for further evolution to contemporary cells.

References

[1] Self Assembly of Structural Subunits. Symposium IV-4, Seventh International Congress of Biochemistry, Tokyo, Japan 19–25 August (1967).
[2] C. Ponnamperuma, Sci. J. (1965) 39.
[3] S. W. Fox, Nature 201 (1964) 336.
[4] C. Ponnamperuma, cited in The origins of prebiological systems and of their molecular matrices, S. W. Fox, ed. (Academic Press, New York, 1965) p. 221.
[5] J. Oro, cited in The origins of prebiological systems and of their molecular matrices, S. W. Fox, ed. (Academic Press, New York, 1965) p. 137.
[6] J. P. Ferris and L. E. Orgel, J. Am. Chem. Soc. 88 (1966) 3829.
[7] S. W. Fox, Nature 205 (1965) 328.
[8] S. Akabori, K. Okawa and M. Saito, Bull. Chem. Soc. Japan 29 (1956) 608.
[9] C. N. Matthews and R. E. Moser, Nature 215 (1967) 1230.
[10] S. W. Fox, cited in Evolving genes and proteins, V. L. Bryson and H. J. Vogel, eds. (Academic Press, New York, 1965) p. 359.
[11] T. Hayakawa, C. R. Windsor, and S. W. Fox, Arch. Biochem. Biophys. 118 (1967) 265.
[12] H. Borsook and H. M. Huffman, cited in Chemistry of the amino acids and proteins, C. L. A. Schmidt, ed. (Charles C Thomas, Springfield, Ill. 1944) p. 822.
[13] M. Dixon and E. C. Webb, Enzymes (Academic Press, New York, 1957) p. 668.
[14] S. W. Fox, Am. Scientist 44 (1956) 347.
[15] G. Steinman, Arch. Biochem. Biophys. 119 (1967) 76.

[16] S. W. Fox and K. Harada, J. Am. Chem. Soc. *82* (1960) 3745.

[17] A. Vegotsky and S. W. Fox, Federation Proc. *18* (1959) 343.

[18] K. Harada and S. W. Fox, cited in The origins of prebiological systems and of their molecular matrices, S. W. Fox, ed. (Academic Press, New York, 1965) p. 289.

[19] S. W. Fox, K. Harada, K. R. Woods, and C. R. Windsor, Arch. Biochem. Biophys. *102* (1963) 439.

[20] S. W. Fox and K. Harada, Federation Proc. *22* (1963) 479.

[21] S. W. Fox and T. Nakashima, Biochim. Biophys. Acta *140* (1967) 155.

[22] R. J. Block and S. Keller, Contrib. Boyce Thompson Inst. *20* (1960) 385.

[23] S. W. Fox and C. T. Wang, Science *160* (1968) 547.

[24] M. Calvin, Am. Scientist *44* (1956) 248.

[25] S. W. Fox, Am. Naturalist *87* (1953) 253.

[26] S. W. Fox, K. Harada, and D. L. Rohlfing, cited in Polyamino acids, polypeptides, and proteins, M. Stahmann, ed. (Univ. of Wisconsin Press, Madison, 1962) p. 47.

[27] J. Noguchi and T. Saito, cited in Polyamino acids, polypeptides, and proteins, M. Stahmann, ed. (Univ. of Wisconsin Press, Madison, 1962) p. 313.

[28] V. R. Usdin, M. A. Mitz and J. Killos, Arch. Biochem. Biophys. *122* (1967) 258.

[29] D. L. Rohlfing and S. W. Fox, Arch. Biochem. Biophys. *118* (1967) 122.

[30] D. L. Rohlfing and S. W. Fox, Arch. Biochem. Biophys. *118* (1967) 127.

[31] S. W. Fox and G. Krampitz, Nature *203* (1964) 1362.

[32] S. W. Fox, cited in The origins of prebiological systems and of their molecular matrices, S. W. Fox, ed. (Academic Press, New York, 1965) p. 361.

[33] D. Durant and S. W. Fox, Federation Proc. *25* (1966) 342.

[34] T. Oshima, Federation Proc. *26* (1967) 451.

[35] G. Krampitz and H. Hardebeck, Naturwissenschaften *53* (1966) 81.

[36] D. L. Rohlfing, Arch. Biochem. Biophys. *118* (1967) 468.

[37] G. Krampitz, S. Diehl, and T. Nakashima, Naturwissenschaften *54* (1967) 516.

[38] A. E. Needham, The uniqueness of biological materials (Pergamon Press, Oxford, 1965).

[39] R. S. Alcock, Physiol. Rev. *16* (1936) 3.

[40] A. I. Oparin, The origin of life on the Earth, 3rd ed. (Oliver and Boyd, Edinburgh, 1957).

[41] G. Wald, Sci. Am., *191* (1954) 44.

[42] J. Lederberg, cited in Current topics in developmental biology, *1*, A. D. Moscona and A. Monroy, eds. (Academic Press, New York, 1966) p. ix.

[43] S. W. Fox, R. J. McCauley, and A. Wood, Comp. Biochem. Physiol. *20* (1967) 773.

[44] A. E. Smith and F. T. Bellware, Science *152* (1966) 362.

[45] S. W. Fox and S. Yuyama, Ann. N.Y. Acad. Sci. *108* (1963) 487.

[46] S. W. Fox and S. Yuyama, J. Bacteriol. *85* (1963) 279.

[47] R. G. E. Murray, cited in The bacteria, Vol. 1, I. C. Gunsalus and R. Y. Stanier, eds. (Academic Press, New York, 1960) p. 55.

[48] D. B. Wetlaufer, Advan. Protein Chem. *17* (1962) 303.

[49] S. W. Fox, R. J. McCauley, T. Fukushima, C. R. Windsor and P. O'B. Montgomery, Federation Proc. *26* (1967) 749.

[50] T. Waehneldt and S. W. Fox, Biochim. Biophys. Acta *134* (1967) 1.

[51] C. Ponnamperuma, cited in The origins of prebiological systems and of their molecular matrices, S. W. Fox, ed. (Academic Press, New York, 1965) p. 221.

[52] A. Schwartz and S. W. Fox, Biochim. Biophys. Acta *134* (1967) 9.

[53] M. Calvin, Bull. Am. Inst. Biol. Sci. *12* (1962) 29.

[54] J. B. S. Haldane, New Biol. *16* (1954) 12.

[55] N. W. Pirie, New Biol. *16* (1954) 41.

[56] J. W. S. Pringle, New Biol. *16* (1954) 54.

[57] N. H. Horowitz, Proc. Natl. Acad. Sci. U.S. *31* (1945) 153.
[58] C. B. Van Niel, cited in The microbe's contribution to biology, A. J. Kluyver and
      C. B. Van Niel, eds. (Academic Press, New York, 1956) p. 155.
[59] S. W. Fox and S. Yuyama, Comp. Biochem. Physiol. *11* (1964) 317.

NON-ENZYMATIC SYNTHESIS OF BIOLOGICALLY PERTINENT PEPTIDES

GARY STEINMAN

*Department of Biochemistry,
Pennsylvania State University, University Park, Pennsylvania, U.S.A.*

Peptide synthesis, like other condensation reactions of biological interest, involves a dehydration process. The union of two amino acids to form a peptide bond yields a molecule of water as a side product:

$$NH_2CHR_1COOH + NH_2CHR_2COOH \rightarrow$$
$$NH_2CHR_1CO-NHCHR_2COOH + H_2O.$$

Such a reaction is thermodynamically unfavorable. A number of methods have been investigated to show how this problem may have been overcome during prebiological times. One approach involves the use of anhydrous, high temperature conditions, as Dr. Fox discusses elsewhere in this volume. Such a method provides a means for supplying the energy necessary for the desired reaction to take place. The high temperature and anhydrous conditions lead to the evaporation of the water side product.

Another possibility for promoting peptide synthesis would be to couple this reaction to the concurrent hydrolysis of another compound. It is this method that I wish to discuss here.

The hypothetical model of prebiotic events that we have employed operates under a general set of restrictions:

1. The reaction system employs simple, relatively stable reactants. This enhances the plausibility of such a model. If the reactants were too complex, it is unlikely that they could have been found on the primitive Earth.

2. The reactions are carried out at moderate temperatures. It is known that biocompounds become very unstable at elevated temperatures [1].

3. An aqueous environment is also employed. Studies designed to duplicate events that could have led up to the origin of living systems should be as "life-like" as possible. Water has been included in our reaction system because the living cell is largely aqueous and it is currently believed that life originated in the seas.

4. Since it is unlikely that the concentration of essential biocompounds

was very high in the primitive oceans, dilute mixtures of reactants were used in these studies.

The work of a number of investigators has indicated that cyanide may well have been one of the key intermediates during prebiological chemical evolution. In particular, the various sparking experiments carried on in "primitive" atmospheres have shown that large amounts of HCN are produced [2]. Furthermore, when aqueous mixtures of ammonium cyanide were heated, both amino acids and peptides were produced [3]. The latter experiments would suggest that the cyanide has served both as the source of building materials for the amino acids themselves and as the condensing agent linking the amino acids together

$$HCN + H_2O \rightarrow HCONH_2.$$

Therefore, a good starting point for seeking out a primitive type of condensing agent would be to consider the chemistry of nitriles.

Various simple nitriles were examined. In particular, the cyanamides (RNHCN) were evaluated for their potential to serve as dehydrating condensing agents in aqueous solutions [4]. Best results were achieved with dicyanamide ($NH(CN)_2$), with the largest yields occurring under acidic conditions [5]. That cyanamides could have been found on the primitive Earth has indicated by their synthesis following the UV irradiation of an aqueous ammonium cyanide solution as well as the sparking of "primitive atmospheres" [6]. Under neutral or basic conditions, the dicyanamide was found to be very stable towards self-hydrolysis [7]. However, upon depression of the pH, the hydrolysis of the cyanamide was accelerated in a similar fashion to its increased reactivity towards peptide synthesis. Thus, dicyanamide can be classified as a normally stable reactant in accordance with restrictions set up for our model.

The mechanism of dicyanamide-mediated peptide synthesis has been worked out in detail [8] (see fig. 1). A restricting factor in this sytem is the amine-catalyzed dimerization of the condensing agent to a nonreactive species (compound IV). Cyanurea, compound VII, is produced as a side-product, serving as the "sink" in which the water removed to form the peptide bond has been isolated. The identification of rearranged ureas (compound VI) strongly supports the contention that an isourea (compound V) probably serves as the intermediate to the reaction. It was found that if one slowly adds the dicyanamide to the reaction system the dimerization of the condensing agent is minimized and peptide synthesis is maximized. This is analogous to events that may have taken place under primitive conditions: the dicyana-

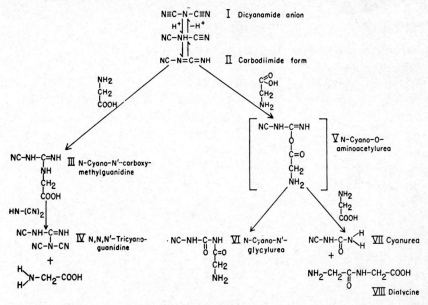

Fig. 1. Summary of the reactions occurring during diglycine synthesis in aqueous solutions of glycine, sodium dicyanamide and HCl [8].

mide could have been synthesized in the upper atmosphere through sparking and ultraviolet irradiation, followed by its being washed by rainwater into the primitive oceans, there to react with amino acids.

Now that a system was available for synthesizing peptides under possible primitive Earth conditions, it was next decided to consider the general nature of peptide synthesis. The first question that was posed was to determine whether amino acid association is random or nonrandom. If the process were totally random, it can be readily shown that the accumulation of specifically sequenced polypeptides would be highly improbable. On the other hand, a nonrandom process would be useful for the synthesis of sequenced polypeptides in the absence of directing enzymes or nucleic acids. A number of amino acid pairs were investigated using dicyanamide. The free association of amino acids turned out to be a nonrandom process, in which certain sequences are preferred over others [9, 10]. (In considering several contemporary proteins whose sequences have already been worked out, similar trends in peptide preferences were noted.) Such trends were observed both in free solution and in mixtures where one of the amino acid reactants was bound to polystyrene resin. The latter could be taken to represent a model of surface phenomena such as occur on membranes in cells.

This line of investigation was continued to determine if other factors inherent in amino acids or found in the environment could have acted to constrain the association of amino acids. The classical picture of enzyme-substrate interactions visualizes the substrate serving as a "key" and the enzyme as a "lock" [11]. The specificity of particular enzymes for only certain keys could account for selective enzyme–substrate interactions based upon structural factors. One possibility would be that in prebiological times the substrates were initially available. These substrate molecules could have served as templates around which the amino acids, later to become parts of active enzymes, arranged themselves in a three-dimensionally specific manner. To test out this possibility, the dimerization of aspartic acid and serine, two residues found at the active site of various hydrolytic enzymes, was studied [10]. It was found that in the presence of a known substrate of chymotrysin, a hydrolytic enzyme, the association of aspartic acid and serine in the manner found at the active site of the enzyme was enhanced over that in the absence of the substrate.

Such environmental factors as pH and ultraviolet light have also been investigated as constraining factors. In the case of pH, it was found that the pK of the reacting amino acids could dictate which residues would be the most reactive [9]. This is probably due to the fact that an undissociated carboxyl group is required in the reaction. It was also found that those amino acids which absorb ultraviolet light become more reactive in dicyanamide-mediated peptide synthesis [12]. This effect is now being investigated with polarized UV light to determine if such a phenomenon could have aided in the establishment of stereo-uniqueness in nature.

The results already discussed indicate that properties inherent in the amino acids, as influenced by factors in the environment, could serve to dictate the sequences that would result in peptide synthesis. In other words, the primary sequence of the resulting peptides is directly influenced by neighbor–neighbor interactions. Studies are now in progress to determine what effect the secondary structure of larger peptides has on sequence generation (fig. 2). In particular, the nonapeptide hormones, oxytocin and vasopressin, are being partially degraded by enzymes. The reassociation of units, as determined by the secondary structure of the peptide fragments, is being investigated for a modifying influence on the ultimate sequences that will result. It is hoped that the residue sequence found in the native hormones will be the preferred one. (For a theoretical treatment, see [13]).

These results suggest that proteins could have predated nucleid acids in the evolution which ultimately led to the appearance of living systems. In this

Vasopressin

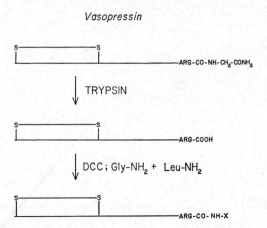

Fig. 2. Use of vasopressin in studying the influence of secondary structure on peptide sequence generation. The C-terminal glycineamide residue is first removed enzymatically. The competition of glycineamide and leucineamide for the available site is examined using carbodiimide-mediated peptide bond synthesis.

way, those sequences which have proved to be of value to evolving biodynamic systems through their utilitarian interactions with environmental factors would be those which would be retained by the organism. The sequences which would be generated and retained would then be translated into polynucleotide sequences. This is the reverse of the manner in which living organisms now translate information (see fig. 3). As is commonly known, organisms now take the information found in nucleic acids and ultimately translate it into protein sequences. The model suggested here would visualize the appearance of sequenced polypeptides first, followed by the storage of the developed information into nucleic acids. This would be better than a "trial-and-error" system which would be the case if the nucleic acids appeared first. There is nothing to suggest that those sequences which may be generated in polynucleotides would subsequently have value in the development of biodynamic systems.

If this model is correct, it is necessary to show how the information generated in certain sequences would then be duplicated in subsequent polypeptides. First of all, it was shown that in the presence of one polypeptide, the synthesis of further polymers is catalyzed [10, 12]. In other words, the process is autocatalytic, a characteristic which is shared with nucleic acid replication. Furthermore, the process appears to be residue-selective [12]. When a poly-L-leucine template was added to a leucine dimerizing system, the union of two

Fig. 3.   Hypothetical scheme for the generation and evolution of biological information.

D-leucine monomers was influenced to a different degree than that of two L-leucine monomers.

Lastly, it should be noted that nearly all proteins found in nature today have their monomeric residues linked by α-peptide bonds exclusively. Some amino acids have reactive side chains. If there were no constraining factor in the association of amino acids, these latter side chains could also serve as sites for the generation of peptide bonds. However, such side chain branching is rarely seen in nature today. It has been found that if one first synthesizes a polymer of amino acids with nonreactive side chains, whereby the α-peptide bond would be the only one that could be produced, and subsequently these side chains are modified to reactive species by appropriate synthetic methods (which have been demonstrated in the laboratory), polymers of amino acids with reactive side chains specifically linked by α-peptide bonds would result [14]. On the other hand, it is difficult to visualize how the 3′–5′ phosphodiester linkage found in nucleic acids could have been specifically generated under prebiotic conditions.

From these results it can be seen that the non-enzymatic generation of biologically pertinent peptides could well have been a highly probable primordial event. The sequences that would be produced would apparently result from information inherent in the reacting units. In this sense, information may be defined as that which sets or reflects a pattern of organization. When one considers all of the experiments that have been carried out in the area of prebiological chemical evolution, it would seem that such preferential interactions occur at many levels of biological order. Overall, the appearance of living systems may, in fact, appear to have been predestined on the basis of the physical and chemical characteristics found in the initial units from which the process began. From this, a biochemical predestination can be hypothesized. (These ideas are developed in greater detail in a monograph published recently [15].) The results now suggest that it is not really necessary to identify which particular synthetic method or group of reaction conditions actually reproduces the exact conditions found on the primitive Earth. Rather, it now seems that the synthesis of biologically significant systems is one which may actually have been difficult to prevent. In other words, the systems found in nature today are quite possibly the result of highly probable events rather than chance occurrences. To extend this idea further, it may be proposed that if any planet exhibited characteristics similar to that found on the primitive Earth, the evolution of Earth-like living systems, in their most general characteristics, would be quite likely. The implications of this to the space program are evident.

Summary. Using dicyanamide, a plausible prebiotic condensing agent, to promote the synthesis of peptides, it has been found that:

1. the association of amino acids by peptide bond coupling is a nonrandom process;

2. the sequences resulting from such a process show marked similarities to the frequency of these sequences in contemporary proteins;

3. a polypeptide can catalyze and constrain a peptide-synthesizing system;

4. environmental factors, such as solvent pH and UV irradiation, modify the nature of the sequences produced; and

5. the presence of a potential substrate can constrain the peptide-synthesizing system to preferentially produce the sequence of residues found in active sites of enzymes.

These results support the suggestion that a protein-based prebiotic system could have arisen and preceded a nucleic acid-based system, as we know it today. While the particular method used here for peptide synthesis may or may not represent the predominant mode active in prebiological times, the results have shown that simple biomonomers possess inherent organizational characteristics which could have effectively aided the course of prebiogenesis.

Acknowledgements. These studies were supported in part by the National Science Foundation, the National Aeronautics and Space Administration, and the College of Science, Pennsylvania State University.

References

[1] P. H. Abelson, Geol. Soc. Am. Mem. Vol. II 67 (1957) 87; J. R. Vallentyne, in The origins of prebiological systems and of their molecular matrices, S. W. Fox, ed. (Academic Press, New York, 1965) p. 105.

[2] S. L. Miller, Biochim. Biophys. Acta 23 (1957) 480.

[3] C. V. Lowe, M. W. Rees, and R. Markham, Nature 199 (1963) 219; M. Labadie, R. Jensen, and E. Neuzil, Bull. Soc. Chim. Biol. 49 (1967) 46.

[4] G. Steinman, R. M. Lemmon and M. Calvin, Proc. Natl. Acad. Sci. U.S. 52 (1964) 27; G. Steinman, R. M. Lemmon, and M. Calvin, Science 147 (1965) 1574; C. Ponnamperuma and E. Peterson, Science 147 (1965) 1573.

[5] G. Steinman, D. H. Kenyon, and M. Calvin, Nature 206 (1965) 707.

[6] A. Schimpl, R. M. Lemmon, and M. Calvin, Science 147 (1965) 149.

[7] G. Steinman and A. Capolupo, Currents in Mod. Biol. 2 (1968) 295.

[8] G. Steinman, D. H. Kenyon, and M. Calvin, Biochim. Biophys. Acta 124 (1966) 339.

[9] G. Steinman, Federation Proc. 26 (1967) 820; G. Steinman, Arch. Biochem. Biophys. 121 (1967) 533.

[10] G. Steinman and M. N. Cole, Proc. Natl. Acad. Sci. U.S. 58 (1967) 735.

[11] E. Fisher, Ber. 27 (1894) 2985.

[12] M. N. Cole, M. S. thesis, The Pennsylvania State University, Univ. Park (1968); G. Steinman and M. N. Cole, Federation Proc. 27 (1968) 765.

[13] H. Pattee, Biophys. J. 1 (1961) 683.

[14] S. Akabori, in Origin of life on the Earth, A. I. Oparin, ed. (Pergamon Press, New York, 1959) p. 189; G. Steinman, Science 154 (1966) 1344.

[15] D. H. Kenyon and G. Steinman, Biochemical predestination (McGraw-Hill, New York, 1969).

FORMATION OF ORGANIC POLYMERS ON
INORGANIC TEMPLATES*

EGON T. DEGENS and JOHANN MATHEJA

*Department of Chemistry, Woods Hole Oceanographic Institution,
Woods Hole, Massachusetts, U.S.A.*

The chemical synthesis of a number of biologically interesting polymers has been successful; this particularly concerns the formation of peptides. A recent account on analytical methods and strategies employed in such work has been given by Bodanszky and Bodanszky [1]. Present techniques of peptide synthesis involve carboxyl activation by, for example, acid chlorides, mixed anhydrides or active esters [2–6]:

benzyl ester   benzoyl (Curtius)   *p* toluene–sulfonyl (Fischer)

phtalyl (Wieland)   benzyl-oxy-carbonyl (Bergmann & Zervas)   tert.-butyl-oxy-carbonyl (Merrifield)

to produce a peptide bond:

peptide bond units

Functional groups not participating in the formation of the amide bond can be rendered inactive by so-called protective groups of which the benzyl-oxycarbonyl group is frequently used [2–8].

* This work was sponsored by the National Aeronautics and Space Administration (NASA-22-014-001) and by a grant from the Petroleum Research Fund administered by the American Chemical Society (PRF-1943-A2). Grateful acknowledgement is hereby made to NASA and the donors of said fund. Contribution No. 2249 from the Woods Hole Oceanographic Institution.

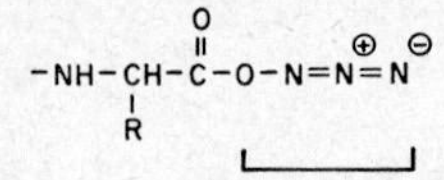

$$-NH-CH-\overset{\overset{O}{\|}}{C}-O-N=\overset{\oplus}{N}=\overset{\ominus}{N}$$
$$\underset{R}{|}$$

acid azide ester
(Curtius)

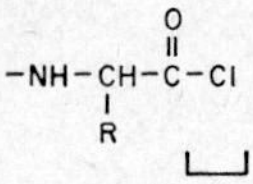

$$-NH-CH-\overset{\overset{O}{\|}}{C}-Cl$$
$$\underset{R}{|}$$

acid chloride
(Fischer)

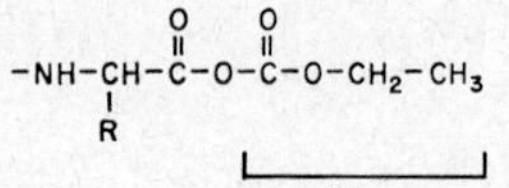

$$-NH-CH-\overset{\overset{O}{\|}}{C}-O-\overset{\overset{O}{\|}}{C}-O-CH_2-CH_3$$
$$\underset{R}{|}$$

ethyl formate
(Wieland)

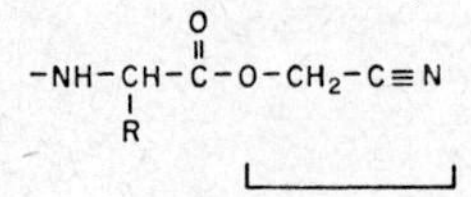

$$-NH-CH-\overset{\overset{O}{\|}}{C}-O-CH_2-C\equiv N$$
$$\underset{R}{|}$$

cyano-methyl ester
(Schwyzer)

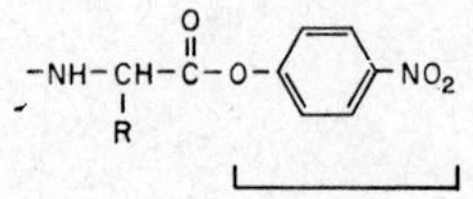

$$-NH-CH-\overset{\overset{O}{\|}}{C}-O-\langle\ \rangle-NO_2$$
$$\underset{R}{|}$$

nitro-phenyl-ester
(Bodanszky)

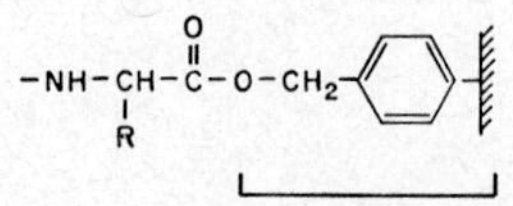

$$-NH-CH-\overset{\overset{O}{\|}}{C}-O-CH_2-\langle\ \rangle$$
$$\underset{R}{|}$$

methyl resin ester
(Merrifield)

An elegant strategy in peptide synthesis is employed by Merrifield [6]. The terminal-COOH group of the amino acid is attached to a styrene resin insoluble in the reaction chamber. This kind of protection is the reason to refer to this method as solid phase peptide synthesis. Studies of this type are principally aimed to yield biologically active peptides and in doing so to obtain independent proof on the correctness of a peptide structuıe inferred from degradation experiments. Examples for a successful synthesis are insulin by Zahn [9], oxytocin and vasopressin by Du Vigneaud [10], or the remarkable story of the adreno-cortico-tropic hormone: ACTH [11,12].

In addition to these well founded chemical interests, there is a biogeochemical reason to study the reaction mechanisms leading to the formation of organic polymers. Biogeochemists are, inter alia, concerned with the origin of proteins and other polymeric building blocks of life during the primordial stage of the earth. From their point of view it is reasonable to assume that mineral matrices participate in the synthesis of organic polymers; they can introduce carboxyl activation and will also function as solid phase protection groups.

To substantiate this supposition, the behavior of amino acids, urea, glucose, the bases of the purines and pyrimidines, and carboxylic acids in the presence of mineral templates was investigated. About 60 different minerals comprising representative members of silicates, oxides, phosphates, carbonates, and sulfates were tested. The selection of minerals was based so as to cover a wide range of crystallographically and chemically different materials [13, 14]. Although we will draw on these results in the present article, the data reported in here are principally confined to polymerization and catalysis experiments involving two representative members of the clay mineral family, the:

montmorillonite: $[(Al, Mg)_2OH_2]^{-1} \cdot [Si_4O_{10}] \cdot [(Na, Ca)\ n\ H_2O]^{+1}$, and
kaolinite: $[Al_4OH_8] \cdot [Si_4O_{10}]$.

Analytical procedures

Contamination. Prior to the experiments on clay mineral-organic interactions, the clays were analyzed for possible organic contaminants of the type studied in the present report. This was necessary because the kaolinite and montmorillonite samples were taken from natural outcrops. Five grams each of kaolinite and montmorillonite (<200 mesh) were subjected to the same analytical treatment in terms of solvent extraction systems, techniques, and instrumental settings that were subsequently used in the actual experiments. The contamination level was found to be below the limit of detection except for the fatty acids where trace amounts were present in the montmorillonite sample. It is concluded that the minerals are essentially free of organic contaminants.

Other precautionary measures were taken by deleting some essential amino acids from the reaction system and by using concentrations far in excess (factor: 10^1–10^4) of those commonly found in fossil clays. In this way a further control on contamination was exercised.

Reaction system. The adsorption and polymerization characteristics as well as the catalytic effects of clay minerals are influenced by a number of parameters of which the physical and chemical nature of the clay, the kind and concentration of organic additives, and the environmental settings, are the most determining ones. To obtain a maximum on information, these three factors were varied within certain limits. Aside of selecting two structurally and chemically different clay minerals, the experiments were performed under wet and dry conditions, low and high temperatures, and different concentration levels for the organic additives. In principal, three sets of experiments were run:

1. 4 ml of a dilute aqueous solution containing 2.5 to 5 μ moles/ml of amino acids, sugars, the bases of the purines and pyrimidines, were added to 2 to 4 g of kaolinite or montmorillonite; the mixture was left in a thermostat in closed glass vessels for 63 hr at 140 °C.

2. 200 ml of a concentrated solution containing 50 to 100 μmoles/ml of 10 amino acids – aspartic acid, serine, glutamic acid, proline, glycine, alanine, valine, phenylalanine, lysine, and histidine – were added to 50 g of kaolinite or montmorillonite; the suspension was left at 80 °C in a closed glass vessel for 7 days; 0.5 ml aliquots were taken at 12 hr intervals. After 7 days, the clay suspension was left at 140 °C under dry conditions; approximately 5 g of sample material each was taken after one and two weeks. The remaining material was left for 3 months at 140 °C.

3. Montmorillonite (1 g), urea, various carboxylic acids, alcohols, and glucose were reacted at 140°C in a closed system for 60 hr. The following analyses were performed: table 1.

TABLE 1

$N_2H-CO-NH_2$ (moles)	Organic additives (moles)	Formula	Name
0.01 (+5 ml H_2O)			blank
0.01	0.05	HO_2CCO_2H	oxalic acid
0.01	0.1	CH_3CO_2H	acetic acid
0.01	0.1	$CH_3CH_2CO_2H$	propionic acid
0.01	0.05	$HO_2CCH_2CH_2CO_2H$	succinic acid
0.1	0.1	$CH_3(CH_2)_6CO_2H$	octanoic acid
0.01	0.05	$HOCH_2CH_2OH$	glycol
0.01	0.05	$HOCH_2CHOHCH_2OH$	glycerol
0.1*	0.033	$C_6H_{12}O_6$	glucose

* 6 g montmorillonite.

Analytical determination. The organic compounds were extracted from the clays following the procedures presented in figure 1. Each listed solvent extraction procedure was performed twice and was followed by a distilled water wash. The silicates were always removed by centrifugation (10,000 rpm; 10 min). The free amino acids, urea and small peptides were analyzed by ion exchange chromatography [15]; the precision of the method is approximately 1% at the 10^{-8} molar level. The bases of the purines and pyrimidines, and the polynucleotides were determined by a combination of molecular sieve and ion exchange techniques [16]. Gel filtration on Sephadex was used as a means of separating classes of organic compounds according to molecular weight and sorption characteristics. The organic solvent extrac-

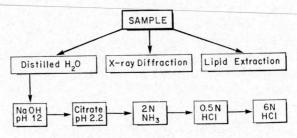

Fig. 1.   Flow diagram of extraction procedures.

table compounds were fractionated on silica gel and were subsequently analyzed by gas chromatography for hydrocarbons, fatty acids, and olefins. Phenols and indoles are presently investigated by paper chromatography. The results and a detailed presentation of the overall analytical methods used in this study will follow. The fractions obtained by the various extraction and separation procedures were tested for their UV and IR sorption characteristics. A flow diagram summarizes the analytical steps taken (fig. 2).

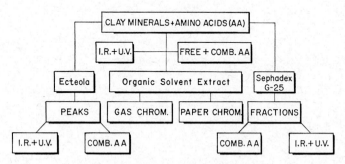

Fig. 2.   Flow diagram of principal analytical steps.

All fractions that were suspected to contain amino acid polymers were hydrolyzed for 22 hr with 6 N HCI. The amino acid spectrum of the hydrolyzate was compared to that of the unhydrolyzed fraction to get a precise estimate on the ratio of free to combined amino acid compounds.

X-ray diffraction analysis was performed before and after the 140°C reaction of the montmorillonite and kaolinite with the amino acids (clay/total amino acid ratio, by weight, approximately 5) in order to determine possible structural alterations. Glycolation and heat combustion experiments at 400, 550, and 650°C were done to assist in the final evaluation of the X-ray data.

Results

Amino acid polymerization (aqueous conditions). Under dry conditions amino acids are taken up or altered by clay minerals in large quantities as indicated by the considerable reduction in free amino acid yield in the first and second distilled water wash of the 140°C kaolinite and montmorillonite specimens.

In contrast, in aqueous suspensions at temperatures below 100°C, amino acids are not picked up in larger quantities. The sorption characteristic of a silicate is a function of the concentration of the solvent, the time of exposure and parameters such as pH or temperature. The Langmuir adsorption equation illustrates these relationships:

$$C = C_\infty \frac{C_0}{C_0 + a},$$

whereby the surface concentration, C, is expressed in moles/cm^2; C_∞ represents the upper boundary limit of the surface concentration and falls in the range of 10^{-10} moles/cm^2 which is equivalent to about 10^{14} molecules per cm^2; C_0 is the concentration of the solvent; and a is a constant, its value depends on solution parameters such as pH, temperature, or the structure of the solvent medium. Diagrams of adsorption relationships for a dilute amino acid solution are shown for kaolinite (fig. 3). This clay mineral exhibits weaker adsorption capacities in comparison to montmorillonite. Kaolinite has no free "inner surfaces", whereas montmorillonite does. The recovery rates of amino acids from highly concentrated solutions, 200 μmoles of each amino acid per 1 g clay, are shown in table 2. The yields of amino acids in the supernatant of montmorillonite and kaolinite suspensions after being left for 7 days at 80°C, show pronounced differences in terms of sorption efficiency and polymerization capacity between the two clay minerals. In general, the basic and acidic amino acids are most drastically affected; but while lysine and

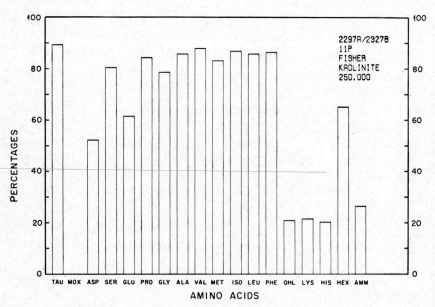

Fig. 3a

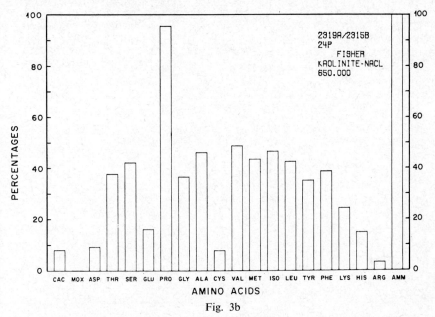

Fig. 3b

Fig. 3.   To assist in the handling and presentation of the analytical data, bar graphs were
routinely drawn by a *Calcomp* plotter which was connected to a GE 225 computer. The
height of the bars indicates the percent recovery of amino acids. Information written in
the right hand corner of the diagrams are for identification of the sample, the procedure,
and the locality.

The upper diagram (a) shows the recovery of free amino acids from a kaolinite which was
subjected to a pressure of 40,000 psi at room temperature for 1 hr; the initial amino acid
input was 10 μ moles/1 g kaolinite. The basic and acidic amino acids are the ones most
drastically reduced. Hydrolysis of the eluate releases most of the missing amino acids.

The sample in the lower diagram (b) was run under identical conditions except that the
amino acids were administered to the kaolinite in a saturated sodium chloride solution.
The recovery rate for free amino acids is significantly lowered. Hydrolysis of the eluate,
however, results in a nearly total recovery of the remaining amino acids.

histidine are simply retained on the clays, aspartic and glutamic acids are
nearly totally recovered upon hydrolysis of the supernatant, kaolinite hereby
being the more effective polymerizer.

It is rather surprising that at temperatures of 80°C, polymers can form in
aqueous solutions. Under these conditions the main polymerization products
are composed of glutamic and aspartic acids whereby the oxidic nature of the
minerals acts as a catalyst. In absence of a mineral suspension, polymerization
does not take place; the amino acids are simply thermally degraded, alanine
being the most stable and serine being the most labile one among the set of
amino acids tested here (table 2); the standard was left for 5 days at 140°C

TABLE 2
Per cent recoveries of amino acids

	Montmorillonite (80°C; 7 days) wet		Kaolinite (80°C; 7 days) wet		Standard* (140°C; 5 days) dry
	free	comb.**	free	comb.**	
Asp	93.7	97.0	3.1	100.0	65.4
Ser	86.9	77.2	98.0	99.6	0.7
Glu	3.0	97.4	1.2	100.1	71.4
Pro	101.3	101.0	100.1	99.9	55.1
Gly	99.6	103.4	99.4	100.0	71.7
Ala	99.9	102.0	89.3	104.3	99.6
Val	102.0	100.5	96.4	94.0	40.8
Phe	98.3	99.6	98.0	99.0	36.9
Lys	10.9	11.0	n.d.	49.3	35.6
His	6.2	6.4	n.d.	47.8	13.8

* 100% recovery in the form of free amino acids was obtained by heating a standard for two days at 100°C under aqueous conditions.
** Per cent recovery of the remaining amino-acids (total-free).

under dry conditions; a standard run at 100°C under wet conditions results in a 100% recovery in the form of free amino acids.

Increase in yield upon hydrolysis of the supernatant suggests that we are principally dealing with a polyglutamic acid product:

polyglutamic acid

with traces of aspartic acid, glycine, alanine, and phenylalanine in a ratio of:

$$Glu:Asp:Gly:Ala:Phe = 100:4:4:2:1.$$

In contrast, kaolinite polymerizes the amino acids in different proportions, i.e.,

$$Glu:Asp:Gly:Ala:Phe = 100:100:1:10:1.$$

Such a polymer is shown in an idealized fashion since no sequence determinations have been made:

Thus, kaolinite catalyses the formation of poly-Asp and -Glu; other amino acids are involved in polymerization only in trace amounts.

The listed amino acids characterize the composition trend of the polymers formed. Under aqueous conditions amino acids such as serine, proline, valine, lysine, and histidine remain inert as far as condensation by means of montmorillonite or kaolinite is concerned. The question arises what is the molecular nature of these products?

We have followed this question, but everybody familiar with this subject matter will attest that sequence determinations of the conformation as to the type of amide bond formed represent a difficult experimental problem. Polymerization may be achieved via the terminal COOH group as is shown schematically for glutamic acid:

$$
\begin{array}{c}
\text{COOH} \\
| \\
\text{CH}_2 \\
| \\
\text{CH}_2 \\
|
\end{array}
$$

CH$_2$–C–N–CH–COOH

H$_2$N–CH–C–N–CH–C–N–CH–C– ...

with CH$_2$–CH$_2$–COOH side chains

Infrared spectroscopy gives certain clues suggesting the presence of peptide bonds. In addition, the fact that under different experimental conditions amino acids, such as glycine, alanine, phenylalanine, and others which do not permit such an amide bond formation, become part of the polymeric products suggests that the peptide linkage type is preferentially formed.

The molecular weights of the amino acid polymers fluctuate between 500 and 5000 with a predominance of around 800 to 1400 MW as ascertained by gel filtration methods. It is thus concluded that the amino acid polymerization products largely represent small peptide chains composed of 7 to 15 amino acid residues.

Amino acid polymerization (dry conditions). The chief objective to raise the temperature to 140°C was to test the thermal stabilities of amino acids and their resulting polymers in a clay-type environment. Two sets of experiments were performed (a) at low amino acid concentration (2.5 μmoles AA per 1 g of clay), and (b) at high amino acid concentration (200 μmoles AA per 1 g of clay).

At temperatures of 140°C the amino acid molecules attached to the silicate surface can be considered a two-dimensional liquid. The melting temperatures for solid amino acids fall in the range of 200 and 300°C. As a solid, amino acids are interwoven in the form of a three-dimensional hydrogen bond network; this is the reason for their high melting temperature. Such a hydrogen bond network no longer persists on silicate surfaces, and the physical state of adsorbed amino acids can thus be considered a liquid rather than a solid. In return, an excellent mobility is maintained for the adsorbed molecules even though only $\frac{1}{5}$ to $\frac{1}{10}$ of all molecules are attached in case of experiment (a) and only $\frac{1}{500}$ to $\frac{1}{1000}$ in case of experiment (b).

By selecting a temperature of 140°C a survey on the catalytical properties of clay minerals was likely to obtain. Furthermore, this temperature is well below the critical temperature which causes "metamorphism" of the sediment material.

The experiments run at 140°C resulted in the decomposition of amino acids as well as the synthesis of peptides and other polymers. We describe a portion of the two experiments to illustrate the processes that are taking place. The first series involves short reaction times (63 hr) and a low concentration of amino acids per unit clay (tables 3 and 4). The second series involves long reaction times (90 days) and large quantities of amino acids per unit clay (tables 5 and 6).

The short range experiments are characterized by a high yield in free amino acids present in the aqueous extracts; for kaolinite the recovery rate is in the order of 10 to 50 per cent, and for montmorillonite it is around 50 per cent.

TABLE 3

Per cent recoveries of amino acids; Kaolinite (63 hrs; 140°C)

| | H_2O | | NaOH | | Citrate | | NH_4OH | | Total yield |
	free	comb.	free	comb.	free	comb.	free	comb.	recovery*
Asp	14.1	20.4	0.8	8.7	0.7	1.5			30.6
Ser	13.0	10.7	0.1	0.9	0.4	0.5			12.1
Glu	3.3	26.4	< 0.1	3.5	< 0.1	0.5			30.4
Pro	17.5	17.2	< 0.1	1.3	< 0.1	< 0.1	not		18.5
Gly	19.3	25.9	0.3	7.4	0.3	1.1	determined		34.4
Ala	45.3	49.2	0.1	7.1	0.3	0.7			57.0
Val	20.4	22.2	0.2	2.0	0.1	0.2			24.4
Iso	15.0	16.1	0.2	1.8	< 0.1	0.2			18.1
Leu	13.2	14.8	0.3	2.4	< 0.1	0.2			17.4
Tyr	9.6	0.6	< 0.1	0.4	< 0.1	< 0.1			1.0
Phe	8.6	9.9	< 0.1	1.9	< 0.1	0.1			11.9
Lys	16.7	13.4	n.d.	1.9	3.8	5.6			20.9
His	12.6	9.5	n.d.	1.6	0.8	0.8			11.9

* The 0.5 N HCl wash and 6 N HCl hydrolysis released only small amounts of lysine and histidine.

TABLE 4

Per cent recoveries of amino acids; Montmorillonite (63 hr; 140 °C)

	H_2O free	comb.	NaOH free	comb.	Citrate free	comb.	NH_4OH free	comb.	Total recovery (%)**
Asp	44.2	43.1	23.4	27.1	2.6	3.4	0.5	↑	74.1
Ser	43.0	44.3	18.3	15.3	1.1	1.5	0.4		61.5
Glu	36.2	47.7	20.6	22.7	0.9	1.0	0.3		71.7
Pro	35.3	32.8	26.7	28.8	1.0	1.3	0.2		63.1
Gly	62.0	61.9	33.9	35.1	1.6	2.8	0.5		100.3
Ala	54.3	52.3	31.8	29.7	1.0	1.3	0.2	*	83.5
Val	59.3	59.7	27.9	27.4	0.5	0.5	0.1		87.7
Iso	53.9	52.8	24.7	24.4	0.3	0.5	0.1		77.8
Leu	52.3	51.1	23.9	23.8	0.4	0.4	0.2		75.5
Tyr	40.8	5.7	18.6	0.4	0.4	0.2	< 0.1		6.4
Phe	45.8	36.8	26.8	26.4	0.6	0.8	< 0.1	↓	64.1
Lys	8.2	5.5	4.2	2.3	6.5	6.4	18.7	18.9	33.1
His	6.1	8.1	1.6	3.2	2.3	4.1	23.7	24.2	39.6

* Sample too small for analysis.
** The 0.5 N HCl wash released only small quantities of lysine and histidine; the 6 N HCl hydrolysis liquor contained larger amounts of unresolved peaks of lysine and histidine.

TABLE 5

Per cent recoveries of amino acids; Kaolinite (90 days; 140 °C)

	H_2O free	comb.	NaOH free	comb.	Citrate free	comb.	NH_4OH* free	comb.	6 N HCl	Total recovery (%)
Asp	< 0.1	0.9	–	0.2	< 0.1	0.1	< 0.1	0.3	1.9	3.4
Ser	< 0.1	0.1	–	< 0.1	< 0.1	< 0.1	< 0.1	< 0.1	< 0.1	0.2
Glu	< 0.1	4.2	–	0.5	< 0.1	0.2	< 0.1	0.3	1.5	6.7
Pro	< 0.1	0.4	–	0.1	< 0.1	0.1	< 0.1	0.1	0.9	1.6
Gly	< 0.1	2.1	< 0.1	0.4	< 0.1	0.4	< 0.1	0.6	2.6	6.1
Ala	< 0.1	2.5	< 0.1	0.5	< 0.1	0.3	< 0.1	0.5	2.8	6.6
Val	–	0.1	–	0.2	< 0.1	0.1	< 0.1	0.2	0.9	1.5
Phe	–	0.5	–	0.2	< 0.1	0.1	< 0.1	0.2	1.1	2.1
Lys	–	0.2	–	0.1	< 0.1	0.1	< 0.1	0.1	0.3	0.8
His	–	0.3	–	< 0.1	< 0.1	< 0.1	< 0.1	< 0.1	0.3	0.8

* The bulk of the organic matter is contained in the NH_4OH wash; evaporation of NH_4OH will result in the formation of a black precipitate which upon 6 N HCl hydrolysis will only release small quantities of amino acids.

The basic amino acids, histidine and lysine, are principally retained on the silicates; but peptides composed of other amino acids have been generated and released to the water. In fig. 4 the increase in amino acid content upon hydrolysis of the first water wash of kaolinite is shown for illustration. The

TABLE 6

Per cent recoveries of amino acids; Montmorillonite (90 days; 140 °C)

	H₂O free	H₂O comb.	NaOH free	NaOH comb.	Citrate free	Citrate comb.	NH₄OH free	NH₄OH comb.	6 N HCl	Total recovery (%)
Asp	9.2	9.6	0.7	1.3	< 0.1	0.3	< 0.1	0.1	< 0.1	11.3
Ser	2.3	2.3	0.2	0.3	< 0.1	0.1	0.1	0.1	< 0.1	2.8
Glu	< 0.1	40.1	< 0.1	3.4	< 0.1	0.3	< 0.1	0.2	< 0.1	44.0
Pro	29.4	31.0	5.1	5.9	2.4	2.4	0.8	0.9	< 0.1	40.2
Gly	25.6	26.2	3.3	4.5	1.3	1.7	0.7	0.9	0.5	33.8
Ala	66.0	68.1	6.8	8.7	1.6	2.5	0.1	1.1	1.2	81.6
Val	32.0	32.2	2.9	3.5	0.1	0.9	< 0.1	0.2	7.1	43.9
Phe	22.3	23.8	2.6	3.3	0.1	0.9	0.2	0.2	5.5	33.7
Lys	3.1	3.1	0.4	n.d.	1.0	1.0	2.8	2.9	69.4	76.8
His	1.3	1.7	0.9	n.d.	0.2	0.2	4.0	4.1	60.6	67.5

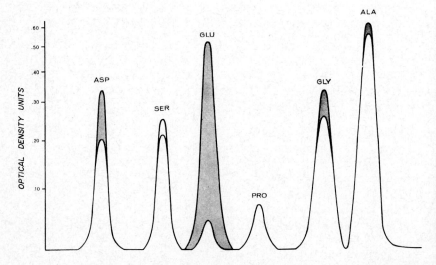

Fig. 4. A segment of the chromatogram of the aqueous extract of kaolinite (63 hr; 140 °C; dry conditions). The shaded area represents the increase in yield upon hydrolysis of the aqueous extract. The reduction in serine is a consequence of its thermal metastability upon acid hydrolysis.

ratio of amino acids in the polymers of the first water wash of kaolinite is as follows:

$$Glu:Asp:Gly:Ala:Val:Leu:Iso:Phe$$
$$23\ :\ 6\ :\ 6\ :\ 4\ :\ 2\ :\ 2\ :\ 1\ :\ 1$$

Different proportions are obtained for the NaOH extract:

Asp: Glu: Ser: Pro: Gly: Ala: Val: Leu: Iso: Phe
8 : 4 : 1 : 1 : 7 : 7 : 2 : 2 : 2 : 2

In the case of montmorillonite, the following relationships are established:

Glu: Ser: His   Asp:Glu: Pro: His
10 : 1 : 2   and   4 : 2 : 2 : 2
(water extract)   (NaOH extract)

The series with the larger amino acid concentrations and longer reaction times shows essentially the same features, except that the level of free amino acids is significantly lowered. As far as the subsequent extracts are concerned i.e., citrate, NH_4OH, 0.5 N HCl, and 6 N HCl hydrolysis for 22 hr, different proportions of amino acids are released, attesting to variations in the specific nature of their attachment to the silicates, and the degree of polymerization and catalysis of the individual amino acids and clay minerals, respectively. For instance, the acidic citrate and ammonia elution releases the basic constituents in greater quantities than the preceding extractions; yet, 6 N HCl hydrolysis is still required to liberate the bulk of lysine and histidine. As far as catalysis is concerned, kaolinite is far more effective than montmorillonite; this also holds true with respect to polymerization effects. For instance, the kaolinite left for 7 to 90 days at 140°C under dry conditions has practically no free amino acids in any of the extracts (table 5). Hydrolysis of the extracts, however, releases large quantities of amino acids. The relationships between the free and combined amino acids in the various kaolinite extracts is shown in fig. 5; the peaks of the free amino acids (shaded area) are enlarged 50-fold.

The low yield in amino acids is unexpected since the initial concentration level was 10^{-2} to 5×10^{-2} moles. Molecular sieve analysis revealed the presence of higher molecular weight compounds with maxima between 1000 and 2000 MW (fig. 6). Some of the fractions collected from the Sephadex column released amino acids upon acid hydrolysis, while others sampled from distinct peaks must represent polymerization products of a different kind.

More than 200 IR spectra were run on the various extracts and fractions. The most outstanding feature is the wide range of OH stretching frequencies with a maximum close to 3300 cm^{-1}. These bands may coincide with the NH stretching bands in the same region. The absorption near 1650 cm^{-1} and 1550 cm^{-1} is assigned to the amide I and II bands (peptide bonds). The presence of aromatic compounds (e.g., phenols), and possibly alcohols is

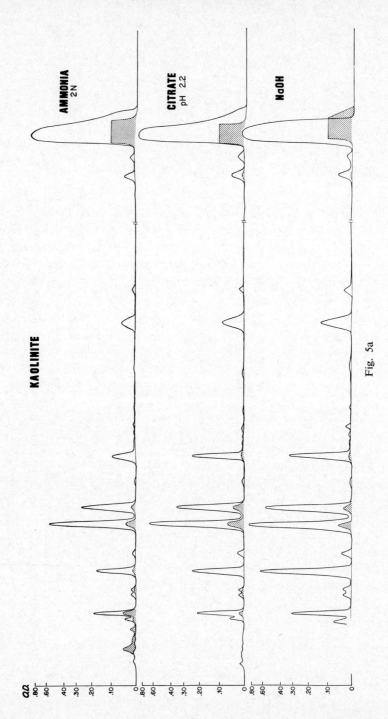

Fig. 5a

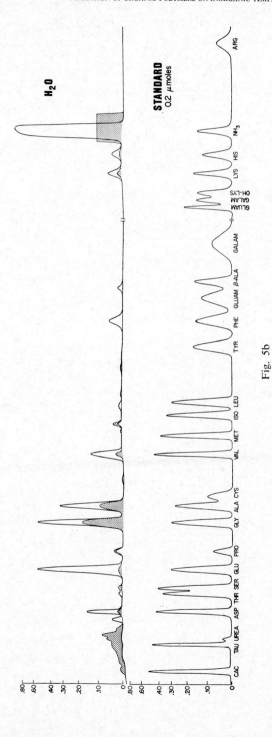

Fig. 5b

Fig. 5. Inasmuch as the recovery rate for the free amino acids was in general less than 0.1 percent for the 90 days kaolinite experiments (table 5), the actual chromatograms of the various extracts both for free (shaded area) and combined amino acids (upper line) are presented. Only $^1/_{50}$ of the volume of the extract that produced the free amino acid peaks produced the spectrum of the combined amino acids upon hydrolysis. A number of amino acids were omitted from the experiments, for instance, threonine, cystine, methionine, isoleucine, leucine, tyrosine, β-alanine, and arginine. Except for β-alanine and threonine none of the others has been found in detectable concentrations.

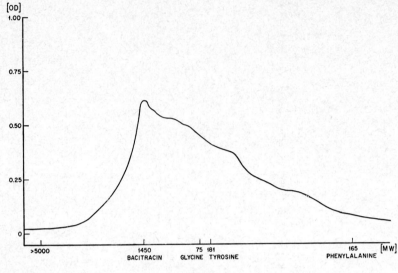

Fig. 6a

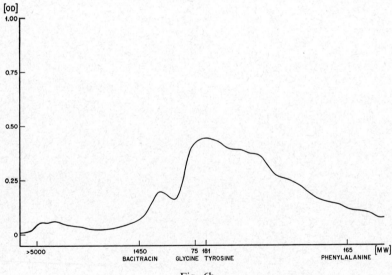

Fig. 6b

indicated by several OH deformations and C−O stretching absorptions in in the region between 1100 cm^{-1} and 1400 cm^{-1}.

Extraction by organic solvents, esterification, and analysis by gas chromatography indicate the presence of fatty acids, hydrocarbons, olefins, and

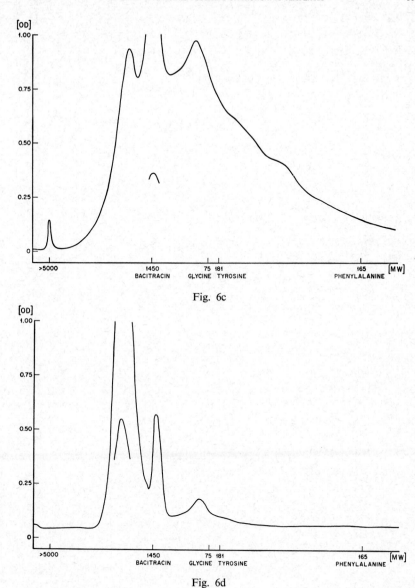

Fig. 6c

Fig. 6d

aromatics. The bulk of the organic solvent extractable material is composed of polar compounds; odd- and even-carbon numbered fatty acids, and hydrocarbons are present in equal concentrations. The ratio of organic solvent extract to initial amino acid input is in the range between 1 to 50 and 1 to

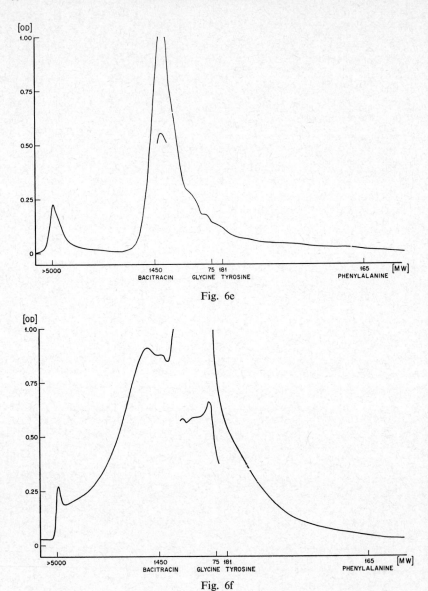

Fig. 6e

Fig. 6f

Fig. 6. A series of representative gel filtration chromatograms run on Sephadex G-25 columns (acetate buffer) are presented. We only include the absorption curves for one wavelength ($\lambda = 280$ mμ); occasionally a cupric acetate solution was injected into the effluent system and a dual wavelength attachment was used to read simultaneously at 280 and 248 mμ. The cupric ion can introduce chelation with alpha and beta amino acids and small peptides. Since the cupric ion has a co-ordination number of four, the empirical

100 and therefore far above the "contamination" level of natural clays. The X-ray diffraction pattern of the kaolinite and montmorillonite samples is not noticeably influenced by reacting the clays at 140°C for 90 days (dry conditions) in the presence of amino acids (200 μmoles AA/1 g clay) (fig. 7). The glycolation and heat combustion characteristics of the clay minerals are not modified by this treatment with amino acids at elevated temperatures. *Peptide synthesis in urea systems.* Experiments have shown [17] the feasibility of urea generation in reaction systems involving H_2, NH_3, and CO at higher temperatures. These experiments contrast previous ones [18] in that no ionizing energies or sparking techniques were applied but simply higher temperatures (130–1050°C); iron meteorite powder was used as a catalyst. It was thus suggestive to study the behavior of urea at elevated temperatures in the presence of mineral catalysts also in view of the fact that the urea molecule is rather stable in aqueous solutions and furthermore can easily be linked to mineral surfaces via hydrogen bonds.

The experiments involving urea, carboxylic acid, alcohols, glucose, and montmorillonite did not produce any free amino acids but remarkable quan-

---

formula of, for instance, copper glycinate is $(NH_2CH_2COO)_2Cu$. In general, however, it was found that the addition of cupric acetate only enhanced the absorption in the lower molecular weight range where the free amino acids appear, but did not produce more absorption peaks in the area where the higher molecular weight compounds occur.

The eluation curve for kaolinite (aqueous conditions; 7 days; 80°C) is shown in chromatogram (a). Hydrolysis of the fraction close to the curve maximum releases principally glutamic and aspartic acids. The tailing is due to the sorption of aromatic compounds to the resin. Chromatogram (b) is the NaOH extract of the kaolinite, run for 90 days at 140°C. Note the appearance of higher molecular weight compounds; glycine, alanine, and glutamic acid are the predominant amino acids in the hydrolysis liquor of the fraction in the proximity of the exclusion volume ($\sim$ 5000 MW). The NH_4OH extract of the same sample [chromatogram (c)] produced a remarkable absorption curve; the fractions from the four principal peaks, however, released only small amounts of amino acids upon hydrolysis. On the basis of IR spectroscopy, polyphenolic compounds probably account for the observed absorption characteristics; the tailing of the curve – due to sorption to the resin – would agree with this statement. The citrate wash of montmorillonite (90 days; 140°C) is shown in chromatogram (d). The fractions under the three principal absorption peaks have not yet been analyzed for combined amino acids. Chromatogram (e) brings the absorption curve of the acetate extract of the montmorillonite-urea-proprionic acid experiment; and chromatogram (f) shows the acetate-extract of the montmorillonite-urea-glycerol experiment. The hydrolyzed fractions under the peaks produce essentially glycine and alanine only; the largest absorption peak in (f) is caused by the presence of glycerol.

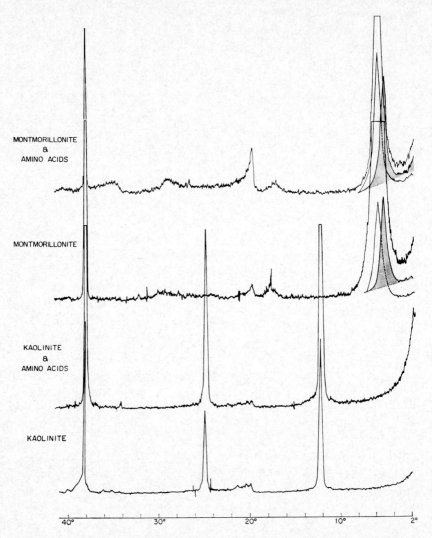

Fig. 7.   X-ray diffraction diagram (CuK$_\alpha$) of montmorillonite and kaolinite. The specimens which were reacted for 90 days at 140 °C under dry conditions in the presence of 200 μmoles amino acids per gram of clay show identical X-ray patterns than the untreated starting material. Glycolation of montmorillonite (shown by the shaded peaks) results in the same degree of expansion, i.e., from ~14 Å to ~17 Å. Heat combustion of kaolinite shows the typical degradation characteristics of kaolinite in both samples. It appears that the amino acids and its polymerization or catalysis products have not structurally modified the original clay minerals. Surface reaction phenomena seem to be responsible for the observed polymerization and catalytic effects.

tities of polymeric compounds, particularly peptides. The amino acid composition in the hydrolysis liquor of some of the runs is shown in fig. 8. The amino acid spectrum of these peptides is rather unique. In the case succinic acid, $HOOC-CH_2-CH_2-COOH$, was administered, aspartic acid, glycine and some alanine were the only amino acids formed and condensed. Glycol, $HO-CH_2-CH_2-OH$, and glycerol, $HO-CH_2-CH-CH_2-OH$, initiated the synthesis of large quantities of glycine and alanine. Several basic amino compounds were also produced; yet, their precise nature could not be determined because the large excess of ammonia from the decomposition of urea interferred with the elution characteristics of the ion exchange column. Alanine, β-alanine, glycine, and valine were formed in the presence of propionic acid, CH_3-CH_2-COOH. The highest yield was obtained upon addition of glucose; again, alanine and glycine were the two most outstanding endmembers. In addition to peptides, other high molecular weight products were generated; their molecular nature is still unknown. In fig. 6 some gel filtration chromatograms are presented.

Purine and pyrimidine experiments. Four bases of the purines and pyrimidines (10 μmoles each) were reacted for 24 hr at 60°C with five different clays (4 g each of kaolinite, halloysite, vermiculite, montmorillonite, and illite). The actual chromatograms of the montmorillonite experiments are presented in fig. 9. The results are as follows:

1. The bases are only moderately retained on the clays except for guanine which shows greater losses; no polymerization is observed.

2. In the presence of sugars (glucose, ribose, xylose, galactose, arabinose; 10 μmoles each), approximately half of the original input in bases remains on the clays; there is no evidence for polymerization.

3. Upon further addition of phosphate (50 mg potassium phosphate), a reaction is achieved as evidence by the disappearance of the bases and the generation of a different chromatographic peak corresponding to higher molecular weight compounds.

4. In the presence of amino acids (18 amino acids; 2.5 μmoles each), phosphates, sugars, and bases, the resulting reaction products are of higher molecular weight than the ones generated in the absence of amino acids. The reaction yields are different for the individual clay minerals. In all instances, however, the addition of phosphates is the most determining factor in the adsorption and polymerization effects. It is noteworthy that hydrolysis with perchloric acid (140°C; 1 hr) of the samples to which phosphate was added will release more than half of the original base input.

Fig. 8a

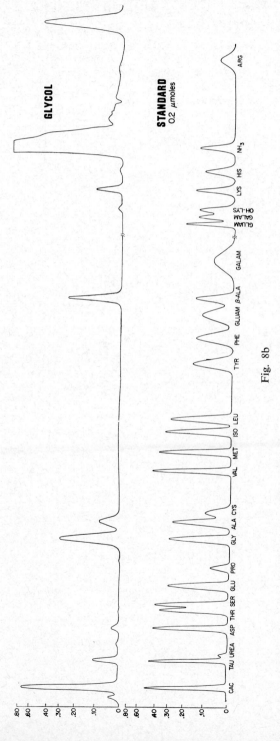

Fig. 8b

Fig. 8. Amino acid chromatograms of hydrolyzed extracts of montmorillonite-urea systems. Analysis for free amino acids produced a straight base line for all samples shown in this figure. Peaks which coincide with the R_f values of cysteic acid, taurine, or urea are still unidentified; they are most likely amines that have been released upon acid hydrolysis. Glycine, alanine, valine, and β-alanine are the most outstanding amino acids. Uncertainty still exists as to the type of the basic amino acids. Each chromatogram represents the yield of $1/25$ of the total sample. In turn, about 5 to 20 μmoles of amino acids have been generated from 10^{-2} moles of urea and 10^{-2} moles of the above listed organic additives. The about 10 times higher amino acid yield for dextrose may simply be the consequence of the about 10-fold larger amount of urea and montmorillonite used in this experiment.

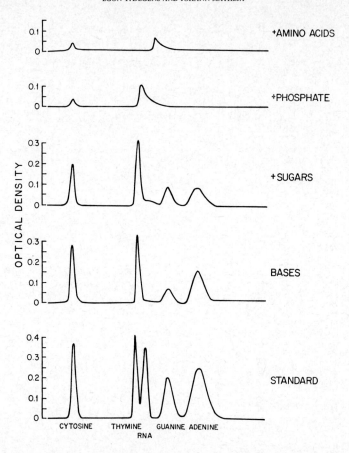

Fig. 9.   Four bases of the purines and pyrimidines (2.5 μmoles/1 g clay) have been reacting with various clays for 24 hr at 60 °C. The chromatograms for the montmorillonite experiments are shown here. A 100 per cent recovery should produce peak heights similar to the peak heights in the standard. The addition of phosphate causes a nearly complete disappearance of the bases. Hydrolysis of the sample residue with perchloric acid for 1 h at 140 °C results in a recovery of more than half of the bases. Since an alkaline extraction has preceded the hydrolysis of the sample, it is tentatively concluded that polymerization of the bases (+ sugars?) has taken place, and that the polymeric products are retained on the clays.

Discussion

The formation of amino acid polymers most likely proceeds via carboxyl activation. The amino groups are either hydrogen bonded to the structural oxygens, or in the case of the basic amino acids, occur as positive ions,

$$Al-OH\overset{\ominus}{..} \ominus \overset{O}{\underset{O}{\text{)}}}C-CH-NH_2 \qquad\qquad Si-O\overset{\ominus}{..}H\overset{\oplus}{-}N\overset{CH_2-CH-COOH}{\underset{NH_2}{\bigg|}}N$$
$$\overset{|}{R}$$

$$Al\overset{\oplus}{..}O\overset{\ominus}{-}\overset{O}{\overset{\|}{C}}-CH-NH_2 \qquad\qquad Si-O\overset{\ominus}{..}H_3\overset{\oplus}{N}-CH_2-CH_2-CH_2-CH_2-CH-COOH$$
$$\overset{|}{R} \qquad\qquad\qquad\qquad\qquad\qquad\qquad\qquad\qquad \overset{|}{NH_2}$$

$$Al\overset{\oplus}{..}O\overset{\ominus}{-}\overset{O}{\overset{\|}{C}}-CH-NH_2$$
$$\overset{|}{CH_2}$$
$$Al-OH\overset{\oplus}{..}\overset{\ominus}{)}\overset{O}{\underset{O}{\text{C}}}-CH_2$$

$$\overset{Si}{\underset{Si}{\text{>}}}O...H_3\overset{\oplus}{N}-\overset{R}{\underset{|}{C}}-COO^{\ominus}$$

and as such are tightly fixed to the silicate surface. This has also been verified by the sorption experiments. There is no need to consider the amino groups as an active element in this system. The carboxyl groups are generally fixed to the positively charged Al-oxy-hydroxy elements by means of ionic bridges. However, it is also conceivable that a direct attachment of the carboxyl groups to the aluminum takes place.

A series of experimental data has been published regarding the isoelectric point of silicates and metal oxides, and the data have been recently reviewed [19]. For silica the isoelectric point is: $pH = 1.8–2.2$; and for aluminum oxides $pH = 8–9$. This suggests that the silica in the clay structure should be negatively charged since the hydroxyls at the surface will release protons:

$$Si-OH \rightarrow Si-O^{\ominus} + H^{\oplus}$$

due to the tight fixation of oxygen to silicon. A release of an OH group from the silicon is thus unlikely.

In contrast to the SiO_4 structural units, aluminum exists in the form of octahedra and is surrounded by six oxygens. In agreement with the high isoelectric point is the statement that aluminum, as a function of its high oxygen co-ordination, either picks up protons or discharges hydroxyls:

$$\underset{Al}{\overset{Al}{\diagdown\diagup}}O ... + H^{\oplus} \rightarrow \underset{Al}{\overset{Al^{\oplus}}{\diagdown\diagup}}OH; \quad Al-OH \rightarrow Al^{\oplus} + OH^{\ominus}.$$

The surface is always positively charged as shown by electrophoresis or the attachment of a gold sol. As evidenced by electron micrographs showing the deposition of negatively charged gold particles at the edges of kaolinite, a

positive charge for the edges has to be assumed [20, 21]. The here proposed
attachments of amino acids to silicate surfaces are therefore coherent with
experimental evidence. Data on the adsorption properties of SiO_2 [22] indicate a surface concentration for Si—OH between 7 and 16 μmoles/m^2; a
simple calculation reveals that each terminal Si—O is actually a Si—OH
hydroxyl.

Differences in the structural composition of kaolinite and montmorillonite account for the observed differences in adsorption, polymerization, and
catalysis of amino acids (fig. 10). The layer sequence for kaolinite is:

silica–alumina–silica–alumina
SiO_4–AlO_6–SiO_4–AlO_6.

Montmorillonite

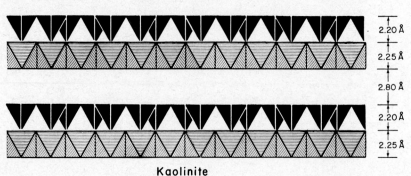

Kaolinite

Fig. 10a

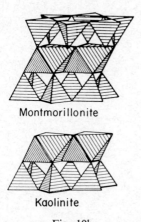

Montmorillonite

Kaolinite

Fig. 10b

Fig. 10. Structure of montmorillonite and kaolinite showing the arrangement and inter-relationships between the silica tetrahedra and the aluminum octahedra. Certain polar compounds such as water or glycol can be introduced into the interlayer space of mont-morillonite and cause expansion of the lattice. The structural dimensions are shown in the 10a pictures and the Si-tetrahedra and Al-octahedra network in the 10b pictures.

The mineral surface is anisotropic, i.e., silica occupies one side, and alumi-num-oxy-hydroxides the other side. In contrast, montmorillonite has no anisotropic surface. The aluminum sheet is incorporated between two silica layers ("sandwich structure") resulting in a layer sequence

[silica–alumina–silica]–silica–alumina–silica–[silica...].

The surface of montmorillonite is only composed of silica. In interlayer position, where oxygens confront each other, metal cations and polar mole-cules are located causing lattice expansion. Montmorillonite is thus capable of forming intercalations with certain organic molecules. It is this structural property of montmorillonite which protects amino acids and explains the high recovery rate for free amino acids even after a reaction time of 3 months at 140°C (table 6). Kaolinite, on the other hand, is the more effective cataly-zer, as a consequence of the nature of its aluminum-oxy-hydroxy surface (fig. 11). As an illustration, we will present a calculation on the adsorption efficiency of clay mineral surfaces.

In our experiments the mineral powder had a particle size smaller than 0.04 mm. The particle radius thus falls in the order of:

$$r = 10^{-3} \text{ to } 10^{-4} \text{ (cm)} = 10^5 \text{ to } 10^4 \text{ (Å)}.$$

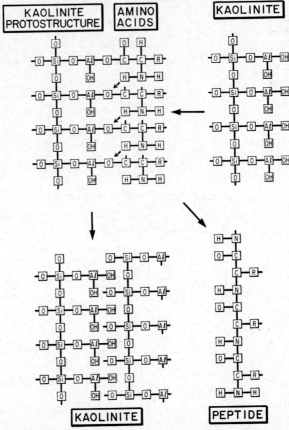

Fig. 11.   Structural relationships in an amino acid-peptide-kaolinite system (schematic). The significance of the aluminum-oxy-hydroxy surface for carboxyl activation and solid phase protection is apparent. Such type of interaction may not only accomplish polymerization of amino acids but will also lead to epitaxial growth of kaolinite.

Clay mineral particles look morphologically like flakes. For convenience we assume a cylindrical form for the clay, whereby the height of the cylinder is approximately $\frac{1}{10}$ of the radius; this kind of height/radius relationship is confirmed by electron micrographs. The surface area of an individual mineral grain amounts to:

$$S = 2\pi r (r + \tfrac{1}{10} r) \approx 10^{10} \ \text{Å}^2$$

and its volume to:

$$V = 2\pi r^2 \, \tfrac{1}{10} r \approx 10^{14} \ \text{Å}^3.$$

In considering the specific density $\varrho \approx 2.6$ g/cm^3, one gram of silicate, having a volume of:

$$0.38 \text{ cm}^3 \approx 4 \times 10^{23} \text{ Å}^3$$

will contain about:

$$4 \times 10^{23}/10^{14} = 4 \times 10^9$$

individual particles which cover a surface area of:

$$4 \times 10^9 \times 10^{10} = 4 \times 10^{19} \text{ Å}^2 = 40 \text{ m}^2/\text{g}.$$

The experiments performed under aqueous conditions for seven days at 80°C (table 2) involved 50 g of clay and 10 amino acids at a concentration level of 10^{-2} moles per AA. The total number of amino acids in the reaction vessel is thus:

$$10 \times 10^{-2} \times 6 \times 10^{23} \approx 10^{23} \text{ molecules.}$$

An adsorbed amino acid requires about 20–30 Å^2 of surface area. In the case of montmorillonite, practically all the glutamic acid has polymerized, and in the case of kaolinite all the acidic amino acids. We have to assume catalysis because the total surface area of the mineral sample:

$$50 \, g \times 4 \times 10^{19} \frac{\text{Å}^2}{g} = 2 \times 10^{21} \text{ Å}^2$$

can accommodate at most:

$$\frac{2 \times 10^{21} \text{ Å}^2}{2\text{–}3 \times 10^{24} \text{ Å}^2} \approx \frac{1}{1000}$$

one thousandths of the amino acid molecules present.

In conclusion, polymerization can only be understood if we assume a flow of molecules across the catalytically effective mineral surface. The result that aspartic and glutamic acids are preferentially polymerized is a consequence of the electrostatic interactions which are responsible for the attachment of molecules to the silicate surface. The expression for adsorption processes can be written [23]:

$$n = \frac{n_\infty}{1 + (n_\infty/n_0 d) \exp - (E/RT)}$$

whereby n represents the surface concentration, n_0 the concentration in the solution, n_∞ the optimal surface concentration, d the thickness of the adsorption layer (≈ 5 Å), and E the adsorption energy. We can assume that the energy is principally of electrostatic nature involving the ionic attachment of amino acids to silicates:

$$E = \frac{e^2}{K \frac{1}{2}d};$$

(e = electrostatic charge unit; K = dielectric constant, d = thickness of adsorption layer) which is proportional to the square of the electrostatic charge units. Glutamic and aspartic acids exhibit two negatively charged oxygens. The adsorption energy is thus four times larger than for a singly charged molecule, and results in an enhanced uptake by the silicates. The fact that the adsorption of amino acids on silicates is of electrostatic nature also explains the preferential adsorption for lysine and histidine. However, no activated carboxyls are produced and polymerization is less pronounced.

Summary

Experimental data are presented on the adsorption, polymerization and catalysis of animo acids on kaolinite and montmorillonite. The following conclusions can be drawn:
1. montmorillonite is a more efficient adsorbent for amino acids than kaolinite;
2. kaolinite is a better catalyst;
3. both minerals introduce polymerization of amino acids under wet and dry conditions;
4. peptides and polyphenolic compounds are the principal condensation products;
5. aspartic and glutamic acids are the chief amino acids to become incorporated in the synthesized peptides;
6. electrostatic interaction and carboxyl activation is the principal driving force in the clay amino acid system;
7. a continuous flow of organic molecules across the catalytically effective mineral surface has to be assumed to understand polymerisation; and
8. a variety of organic molecules including fatty acids, hydrocarbons and phenols can be generated in a clay-amino acid system.

Amino acid polymers can also be produced using urea, carboxylic acids, alcohols and sugars. The organic additives determine the nature of the generated amino acids and the clay minerals promote catalysis and polymerization, and will also function as solid phase protection groups.

The adsorption and polymerization studies of the bases of the purines and pyrimidines on clay minerals indicate that the addition of phosphates is the most determining factor in the retention of the bases; there is indication that polynucleotides do form under these circumstances.

Acknowledgements. We are grateful to Dr. M. Blumer for the gas chromatographic analyses and to Mr. J. Hathaway for the X-ray diffraction analyses. The technical assistance of Miss R. Grant and Mrs. M. Hummon is gratefully acknowledged.

Appendix

Since submission of this manuscript (July 1968) some additional information was obtained. It could be shown that by using the lattice of a polar aluminum silicate, i.e. kaolinite, an asymmetric synthesis of peptides is accomplished and that the polymers containing left-handed amino acids do preferentially form over those containing D-optical isomers [24]. Aside of polymerization and selection of isomers, silicates are capable of catalyzing the reaction:

$$CO_2 + 2 NH_3 \rightleftharpoons H_2N - CO - NH_2 + H_2O.$$

Inasmuch as our experiments indicate [25] that amino acids and urea are in thermodynamic equilibrium and furthermore that amino acids polymerize in the presence of a solid phase boundary, we can write the following generalized equation:

$$n_1CO_2 + n_2NH_3 \rightleftharpoons n_3(H_2N)_2CO + n_4H_2O \rightleftharpoons n_5(AA) \rightleftharpoons n_6(PP) + H_2O.$$

In conclusion, in natural environments and in the presence of minerals a reaction chain is established which starts from simple gases (CO_2 and NH_3) and which may lead via urea and amino acids (AA) to polypeptides (PP). During polymerization of amino acids, peptides containing L-optical isomers are preferentially formed.

In another set of experiments (Harvey and Degens, unpublished) kaolinite (20 g) was reacted with glycerin (2 g) and palmitic acid (2 g) in 100 ml of water at 85 °C for 2 weeks. Crystals of tripalmitin (3 g) and 1, 2-dipalmitin (5 mg) were obtained. We were also successful (Mopper and Degens, unpublished) to synthesize a variety of sugars in a formaldehyde-kaolinite system. It is interesting to note that 2 N H_2SO_4 hydrolysis was needed to release the sugar monomers.

The implications of this kind of research for prebiotic and biochemical evolution have been presented elsewhere [26].

References

[1] M. Bodanszky and A. A. Bodanszky, Am. Sci. 55 (1967) 2.
[2] T. Curtuis, Ber. 35 (1902) 3226.
[3] E. Fischer, Ber. 36 (1903) 2094.
[4] T. Wieland, et al., Ann. 569 (1950) 117.
[5] M. Bergmann and L. Zervas, Ber. 65 (1932) 1192.
[6] R. B. Merrifield, J. Am. Chem. Soc. 86 (1964) 304.
[7] R. Schwyzer, et al., Helv. Chim. Acta 38 (1955) 80.
[8] M. Bodanszky, Nature 175 (1955) 685.
[9] H. Zahn, W. Danho and B. Gutte, Z. Naturforsch. 21b (1966) 763.
[10] V. Du Vigneaud, Harvey Lectures Ser. 50 (1956) 1.
[11] R. Schwyzer, Ann. Rev. Biochem. 33 (1964) 259.
[12] R. Schwyzer and P. Sieber, Helv. Chim. Acta 49 (1966) 134.
[13] E. T. Degens and J. Matheja, Woods Hole Techn. Rep. No. 67–57 (1967) p. 316.
[14] E. T. Degens and J. Matheja, J. Brit. Interplanetary Soc. 21 (1968) 52.
[15] E. T. Degens and D. W. Spencer, Woods Hole Techn. Rep. No. 66–27 (1966) p. 32.
[16] S. Keleman and E. T. Degens, Nature 211 (1966) 857.
[17] R. Hayatsu, M. H. Studier, A. Oda, K. Fuse and E. Anders, Geochim. Cosmochim. Acta 32 (1968) 175.
[18] N. H. Horowitz and S. L. Miller, Fortschr. Chem. Org. Naturstoffe 20 (1962) 423.
[19] G. Parks, Chem. Rev. 65 (1965) 177.
[20] P. A. Thiessen, Elektrochem. 48 (1942) 675.
[21] J. Mering, A. Mathieu-Sicaud and J. Perrin-Bonnet, Compt. Rend. Congr. Geol. Int. Algier, Section 18 (1952) p. 103.
[22] M. M. Egorow, W. J. Kvlividze, V. F. Kisselev and K. G. Krassilnikov, Kolloid-Z. 212 (1966) 126.
[23] J. I. Frenkel, Statistische Physik (Akademie Verlag, Berlin, 1957).
[24] E. T. Degens, J. Matheja and T. A. Jackson, Nature, 227 (1970) 492.
[25] J. Matheja and E. T. Degens, Abstract, Third International Bioph. Congr. Int. Union Pure and Applied Biophys., Cambridge, Mass. Aug. 29–Sept. 3 (1969) 169.
[26] J. Matheja and E. T. Degens, Structural molecular biology of phosphates (Gustav Fischer Verlag, Stuttgart, 1970).

POSSIBLE MECHANISMS FOR PREBIOTIC
PHOSPHORYLATION

J. RABINOWITZ, S. CHANG and CYRIL PONNAMPERUMA

Exobiology Division,
Ames Research Center, NASA, Moffett Field, California, U.S.A.

Phosphoryl compounds must have played as essential a role in the early stages of chemical evolution as they do in living systems today. Of particular importance would have been sugar, nucleoside, and other organic phosphates. In the pH 7–9 region, generally considered appropriate for primitive bodies of water, and in the presence of alkaline earth ions, such mono- and diesters of phosphoric acid are generally soluble and relatively stable to hydrolysis. These properties would have made phosphate esters well suited to serve as precursors of more complex biopolymers in primordial waters. The search for prebiotic phosphorylating agents and the synthesis of phosphate esters under presumably prebiotic conditions, therefore, constitute important areas of research in chemical evolution.

Two general experimental approaches have received attention: dry thermal reactions between inorganic phosphates and organic substrates, and aqueous phosphorylations involving reactive intermediates formed by addition of inorganic orthophosphate to potential prebiotic condensing agents. The latter reactions fall into a mechanistic class developed by synthetic organic chemists, which involve nucleophilic displacement at tetrahedral phosphorus accompanied by heterolysis of a labile bond between phosphorus and an activating group X [1]. Compounds reportedly phosphorylated included glucose, nucleosides, and inorganic orthophosphate, and some of the reagents

used were cyanamide [2] $_y$X = −OC(NH$_2$)=NH$_7$, N-cyanoguanidine [2] [X = −OC(NH$_2$)=NC(NH$_2$)=NH$_7$], cyanogen [3] $_y$X = OC(CN)=NH$_7$, cyanate [3, 4] [X = −OC(OH)=NH], cyanofornamide [3] [X = −OC(CO-NH$_2$)=NH], and cyanoacetylene [5] [X = −OCH=CHCN]. The same mechanism undoubtedly prevailed in thermal phosphorylations of nucleosides with polyphosphoric acid [6] [X = −O(PO$_3$H)$_n$H, $n \geq 1$].

The picture with dry thermal reactions of nucleosides with inorganic orthophosphates [7,8], however, is less clear. Good yields of nucleoside phosphates were obtained at 160°C with dihydrogen orthophosphates. Marked decreases in yields were observed with monohydrogen phosphates, M_2HPO_4, except when one M was ammonium, in which case thermal dissociation to ammonia and dihydrogen phosphate was possible. Practically no phosphorylation took place ($<1\%$) when tribasic phosphates were used [7]. Because, to our knowledge, there were no prior reports of direct phosphorylation of organic hydroxy compounds with inorganic orthophosphates, it was of interest to investigate the course of this reaction.

Two mechanisms or a combination of both may have been operating: direct phosphorylation by a path other than that illustrated above, or partial thermal transformation of orthophosphates to condensed phosphates, the latter serving as the effective phosphorylating agents. In the latter case, the reaction would conform to the equation with N = nucleoside and X = O-$(PO_3^-)_nH$, $n \geq 1$. To test this hypothesis, various orthophosphates were heated in open vials for 2 hr at 160° and analysed for polyphosphates by

TABLE 1

Conversion of orthophosphates to condensed phosphates at 160°C after 2 hr

Orthophosphate used[a]	% of total phosphorus as pyrophosphate[b]	as tripolyphosphate[b]	% Uridine[c] converted to monophosphate
H_3PO_4	30–50	1–5	8.3
$NaH_2PO_4 \cdot H_2O$	5–10	1–5[d]	16.0
$Na_2HPO_4 \cdot 7H_2O$	–	–	0.6
$Na_3PO_4 \cdot 12H_2O$	–	–	0.6
$NH_4H_2PO_4$	10–30	5–10[d]	5.9
$(NH_4)_2HPO_4$	10–30	5–10[d]	13.4
$Ca(H_2PO_4)_2 \cdot H_2O$	30–50	10–30[d]	10.5
$CaHPO_4$	–	–	
KH_2PO_4	–[e]	–	
$Ca_{10}(PO_4)_6(OH)_2$	–[f]	–	
$Ca_3(PO_4)_2$	–	–	0.1

[a] Orthophosphates were used as obtained commercially. No condensed phosphates were detectable prior to heating.
[b] Visual estimates made by comparing the sizes and color intensities of spots with those obtained with standard mixtures having known ortho- and polyphosphate compositions. Less than 1% indicated by –.
[c] Taken from ref. [7].
[d] Polyphosphates more highly condensed than tripolyphosphate were observed at or near the origin.
[e] Pyrophosphate appears after 6 hr at 160°C.
[f] No condensed phosphates appeared after 7 days at 160°C.

paper chromatography [9]. Some results are summarized in table 1 along with yields obtained in phosphorylation of uridine under identical conditions [7].

The conversion of orthophosphates to polyphosphates was most extensive with H_3PO_4 and $Ca(H_2PO_4)_2 \cdot H_2O$, affording pyro, tri, and higher polyphosphates [10]. Significantly, no polyphosphates were detectable with dibasic and tribasic phosphates which previously gave very low yields of nucleotides ($<1\%$). The principal phosphate in igneous and metamorphic rock, hydroxylapatite, prepared according to Miller and Parris [4], gave negative results after prolonged heating. Data in table 1 show that phosphorylating agents, that is polyphosphates, containing the requisite $P-X$ bonds, are formed under conditions where appreciable thermal phosphorylation of nucleosides with inorganic orthophosphates was observed [10].

To obtain quantitative data on the rates and extents of polyphosphate formation, the thermal condensations of $NaH_2{}^{32}PO_4 \cdot H_2O$, $NH_4H_2{}^{32}PO_4$, and $Ca(H_2{}^{32}PO_4)_2 \cdot H_2O$ were studied at several temperatures. Aliquots (10 μl) of solutions of the various orthophosphates (0.09 M $NaH_2{}^{32}PO_4 \cdot H_2O$, 2.12 $\mu C/\mu l$; 0.03 M $NH_4H_2{}^{32}PO_4$, 2.43 $\mu C/\mu l$; 0.08 M $Ca(H_2{}^{32}PO_4)_2 \cdot H_2O$, 0.196 $\mu C/\mu l$) were lyophilized in small vials and heated for various lengths of time at suitable temperatures. After cooling, the solid was dissolved in water (10 μl) and the solution (1 μl) chromatographed on Eastman Chromagram cellulose sheets. All radioactive spots were located with X-ray films and cut out, and the radioactivity measured in a liquid scintillation counter. Chromatography [9] revealed only linear polyphosphates; in no instance were cyclic metaphosphates detected. Some results are presented in table 2.

Data obtained after 0.5 hr at 162 °C and after 6 hr at 126 °C demonstrate that the ease of polyphosphate formation is in the order $Ca(H_2{}^{32}PO_4)_2 \cdot H_2O$ $> NaH_2{}^{32}PO_4 \cdot H_2O > NH_4H_2{}^{32}PO_4$. Notably, the calcium salt is converted to 2.2% pyrophosphate at as low a temperature as 92 °C after 18 hr. In table 3 are qualitative results obtained at 90 °C and 65 °C with unlabeled phosphates. These are the lowest temperatures at which direct thermal conversion of an orthophosphate to pyrophosphate has been observed. Previously, traces of pyrophosphate were detected chromatographically after heating $CaHPO_4 \cdot H_2O$ at 120 °C for 20 hr [11]. Similarly, polyphosphates have been detected after heating LiH_2PO_4 above 145 °C, NaH_2PO_4 above 160 °C, and $NH_4H_2PO_4$ above 140 °C for many hours [12]. In none of these instances, however, were the amounts of polyphosphates determined.

Clearly, inorganic orthophosphates can be converted to polyphosphates at significant rates at temperatures much lower than those previously observed and those required to achieve near-complete polymerization [11–14]. The

TABLE 2

Thermal conversion of ^{32}P-labelled orthophosphates to polyphosphates at several temperatures

$T(°C)$	Orthophosphates[a] used	Time (hr)	% of total radioactivity[b] as					
			Orthophosphate	Pyrophosphate	Triphosphate	Tetraphosphate	Pentaphosphate	Hexa- and higher phosphates
162.0	$Ca(H_2{}^{32}PO_4)_2 \cdot H_2O$	0.1	63.3 ± 2.3	21.7 ± 2.5	9.9 ± 0.4	3.1 ± 0.2	1.3 ± 0.1	0.8 ± 0.0
		0.5	33.6 ± 4.0	25.3 ± 0.9	18.3 ± 1.5	8.5 ± 0.6	4.7 ± 0.7	9.6 ± 0.3
	$NaH_2{}^{32}PO_4 \cdot H_2O$	0.1	97.9	2.1				
		0.5	80.0	17.2	1.9	0.8		
		1.5	78.9	16.8	3.4	1.2	0.7	
	$NH_4H_2{}^{32}PO_4$	0.5	97.5 ± 0.6	2.3 ± 0.3	0.4 ± 0.2			
		1.0	92.3 ± 2.1	6.6 ± 1.6	1.8 ± 0.4			
126.0	$Ca(H_2{}^{32}PO_4)_2 \cdot H_2O$	6.0	82.6 ± 0.9	12.5 ± 0.7	4.8 ± 1.7	0.4 ± 0.2		
	$NaH_2{}^{32}PO_4 \cdot H_2O$	6.0	97.3	2.7				
	$NH_4H_2{}^{32}PO_4$	6.0	99.4 ± 0.2	0.6 ± 0.2				
92.0	$Ca(H_2{}^{32}PO_4)_2 \cdot H_2O$	18.0	97.8	2.2				

[a] Chromatographic analysis of the stock solutions of orthophosphates revealed that less than 0.06% of the total radioactivity was located with R_f values corresponding to those expected for polyphosphates.

[b] Figures including ± signs indicate average values and average deviations for two independent experiments.

TABLE 3

Conversion of orthophosphates to polyphosphates at 90 °C after one week
and at 65 °C after two months

Orthophosphate[a] used	T (°C)	% of total phosphorus[b] as	
		Pyrophosphate	Triphosphate
$Ca(H_2PO_4)_2 \cdot H_2O$	90 (after 1 week)	2–5	2–5
$NaH_2PO_4 \cdot H_2O$		2–5	2–5
$NH_4H_2PO_4$		~1	~1
$Ca(H_2PO_4)_2 \cdot H_2O$	65 (after 2 months)	~1	–
$NaH_2PO_4 \cdot H_2O$		~1	–

[a,b] See table 1.

plausibility of prebiotic polyphosphates produced thermally, however, may be questioned on the same grounds as that of any of the other presumed prebiotic phosphorylating agents. That is, in the presence of calcium ions and in waters having $pH \geq 7$, it is believed that essentially all orthophosphate ions would be precipitated as insoluble hydroxylapatite [4]. Hydroxylapatite prepared according to Miller and Parris [4] is incapable of thermal polymerization at moderate temperatures (see above). However, it is known that hydroxylapatites obtained from different preparations have variable compositions with Ca/P ratios ranging from 1.33 to 2.0 [15]. When the ratio is 1.33, as in "octacalcium phosphate," $Ca_4H(PO_4)_3 \cdot 2H_2O$, thermal transformation to pyrophosphate and hydroxylapatites having higher Ca/P ratios was observed at temperatures as low as 150 °C. Furthermore, if primordial oceans ever contained low concentrations of alkaline earth ions or if localized geological phenomena permitted the existence of bodies of water having lower pH values (<7), precipitation of various hydrogen and dihydrogen phosphates and their subsequent thermal polymerization would have been plausible. The simplicity of the reaction would then make thermal polymerization of inorganic orthophosphates at moderate temperatures very attractive as a general source of polyphosphates, potential [4, 6, 16] phosphorylating and condensing agents in primordial syntheses in aqueous [17] or dry thermal conditions.

To assess the relative effectiveness of orthophosphate and polyphosphates as phosphorylating agents, uridine was heated with $NaH_2PO_4 \cdot H_2O$ at 162 °C and 126 °C for times during which quantitative estimates could be made of the amount of inorganic polyphosphate formed (see table 2). The expected products [7, 8] were separated chromatographically and estimated quantitatively by UV analysis. The yield of uridine phosphates was compared

TABLE 4

Conversion of uridine to uridine phosphates by heating the nucleoside with various inorganic phosphates

T (°C)	Time (hr)	Phosphate[a,b] used	% Uridine converted to phosphates[c,d]	% of total phosphorus as[e]	
				Pyrophosphate	Tri- and higher phosphates
162.0	0.1	$NaH_2PO_4 \cdot H_2O$	2.0 ± 0.4[f]	<3[g]	0
		$NaH_2PO_4 \cdot H_2O$ (heated)	5.5 ± 1.0[f]	(25)	(7)
		$Na_5P_3O_{10}$	15.4 ± 4.2[f]		(>95)
126.0	6.0	$NaH_2PO_4 \cdot H_2O$	2.9	<3[g]	0
		$NaH_2PO_4 \cdot H_2O$ (heated)	7.4	(25)	(7)
		$Na_2H_2P_2O_7$	29.8	($>95\%$)	

a All phosphates used as obtained commercially. The orthophosphate contained no detectable ($<1\%$) polyphosphates, the pyrophosphate about 5% total of ortho- and triphosphate, and the triphosphate about 5% total of mono-, di-, and tetraphosphate. The "heated" phosphate was obtained by heating $NaH_2PO_4 \cdot H_2O$ at 162°C for 5 hr. On the basis of studies with $NaH_2{}^{32}PO_4 \cdot H_2O$, the % of total phosphorus expected as various phosphates was: 68% mono-, 25% di-, 4% tri-, 2% tetra-, and 1% pentaphosphate.

b Equimolar amounts of phosphate and uridine were used except in reactions involving "heated" monophosphate, in which case the weight of phosphate was the same as that of the monophosphate in the parallel experiment.

c Products separated on Brinkmann cellulose F TLC plates in isobutyric acid/0.5 N NH_3 (60/36) and eluted from the plates with tris HCl buffer (pH 7).

d Measured as % of total absorption at 260 mμ recovered from chromatograms.

e Figures in parentheses refer to initial compositions.

f Average and average deviation of two independent experiments.

g Based on table 2.

with that obtained when uridine was heated under the same conditions with $Na_2H_2P_2O_7$, $Na_5P_3O_{10}$, and with an orthophosphate-polyphosphate mixture. The results in table 4 indicate that polyphosphates are indeed more effective phosphorylating agents. The amount of phosphorylation observed in the orthophosphate experiments may have resulted from reaction of uridine with the small amount of polyphosphates expected to have formed during the period of heating.

These findings do not eliminate the possibility that $NaH_2PO_4 \cdot H_2O$ also phosphorylates directly by a mechanism other than that illustrated above. They do indicate, however, that sodium polyphosphates formed in these reactions will phosphorylate the nucleoside much faster than the original orthophosphate even if an additional mechanism operates. Thus, polyphosphates are certainly intermediates in the phosphorylation of nucleosides at 160 °C with NaH_2PO_4. In addition, they have been shown to be effective in phosphorylation of adenosine in aqueous solutions [17]. Further investigation of their intermediacy in phosphorylations involving other orthophosphates are under way, as are inquiries into their efficacy in promoting various condensation and polymerization reactions in aqueous solutions.

References

[1] D. M. Brown, Advances in organic chemistry, methods and results, R. A. Raphael, E. C. Taylor and H. Wynberg, eds. (Interscience Publishers, New York, 1963); H. G. Khorana, Some recent developments in the chemistry of phosphate esters of biological interest (John Wiley and Sons, New York, 1961); A. J. Kirby and S. G. Warren, The organic chemistry of phosphorus (Elsevier, New York, 1967).

[2] G. Steinman, R. M. Lemmon and M. Calvin, Proc. Natl. Acad. Sci. U.S. 52 (1964) 27.

[3] R. Lohrmann and L. E. Orgel, Science 161 (1968) 64.

[4] S. L. Miller and M. Parris, Nature 204 (1964) 1248.

[5] J. P. Ferris, Abstract of Paper No. 73 presented at the 156th Meeting of the American Chemical Society, Atlantic City, New Jersey (Sept. 9–13, 1968).

[6] T. V. Waehneldt and S. W. Fox, Biochim. Biophys. Acta 134 (1967) 1.

[7] C. Ponnamperuma and R. Mack, Science 148 (1965) 1221.

[8] J. Skoda and J. Moravek, Tetrahedron Letters 1966 (1966) 4167; J. Moravek and J. Skoda, Collection Czech. Chem. Commun 32 (1967) 206; J. Moravek, Tetrahedron Letters 1967 (1967) 1707; A. Beck, R. Lohrmann and L. E. Orgel, Science 157 (1967) 952.

[9] E. Karl-Kroupa, Anal. Chem. 28 (1956) 1091; R. J. Fuchs and F. W. Czeck, Anal. Chem. 35 (1963) 796.

[10] J. Rabinowitz, S. Chang and C. Ponnamperuma, Nature 218 (1968) 442.

[11] H. Newesely, Monatsh. Chem. 98 (1967) 379.

[12] E. Thilo and H. Grunze, Z. Anorg. Allgem. Chem. 281 (1955) 262.

[13] J. W. Edwards and A. H. Herzog, J. Am. Chem. Soc. 79 (1957) 3647; W. L. Hill, S. B. Hendricks, E. J. Fox and J. G. Cady, Ind. Eng. Chem. 39 (1947) 1667.

[14] J. R. Van Wazer, Phosphorus and its Compounds (Interscience Publishers, New York, 1958) ch. 10.
[15] J. R. Van Wazer, ibid. p. 511.
[16] F. M. Harold, Bacteriol. Rev. *30* (1966) 772.
[17] A. Schwartz and C. Ponnamperuma, Nature *218* (1968) 443; see also presentation by these authors.

PHOSPHORYLATION OF NUCLEOSIDES BY
CONDENSED PHOSPHATES IN AQUEOUS SYSTEMS

ALAN SCHWARTZ* and CYRIL PONNAMPERUMA

*Exobiology Division, National Aeronautics and Space Administration,
Ames Research Center, Moffett Field, California, U.S.A.*

Little is known about the state of phosphorus on the primitive earth. The geological evidence is inconclusive, and open to differences in interpretation – particularly, may I add, on the subject of the composition of the primitive ocean. We are therefore left with the necessity of "projecting backward", as Fritz Lipmann has pointed out, from biological systems to the prebiological [1]. The very widespread occurrence of inorganic polyphosphates in micro-organisms suggests that polyphosphate is a kind of "metabolic fossil", and that primitive micro-organisms may have made much greater use of poly-phosphates as primary energy sources [2]. It is of course, a logical leap, but a tempting one, to suggest that inorganic polyphosphates may have been important energy sources on the prebiological earth as well.

Long chain polyphosphates are produced in the heating of orthophosphate salts at elevated temperatures. More significantly perhaps, from the view-point of prebiological chemistry, the lower members of the series are formed under very mild conditions indeed, as has been reported by Dr. Rabinowitz. Once formed, the linear polyphosphates are quite stable in aqueous solution. The tripolyphosphate ion, for example, has a half-life at room temperature and neutrality which is measured in years [3]. Because of the possible role of water soluble polyphosphate salts as prebiological phosphorylating agents, experiments on the phosphorylation of adenosine were conducted with tetrasodium pyrophosphate, and pentasodium tripolyphosphate, both of which were commercial products. To examine the effect of longer chain polyphosphates, a preparation of Graham's salt was made by melting sodium dihydrogen phosphate at 650 °C, holding the melt at that temperature for 4 hr, and quick cooling. The preparation was shown to be composed primarily of linear polyphosphate chains, with no species smaller than tri-polyphosphate detectable by thin layer chromatography [4]. In a typical reaction, an amount of polyphosphate salt corresponding to 20 mmoles of phosphorus, and 2 mmoles of adenosine were weighed into a 50 ml round

* Present address: Department of Exobiology, University of Nijmegen, Nijmegen, The Netherlands.

bottom flask. The mixture was dissolved with heat in 20 ml of water, and the solution was refluxed for 4–6 hr. The hot solution was then diluted with water, cooled, and run directly onto a column of Dowex 1 × 2 formate. Elution was begun immediately with 0.2 N formic acid. Fig. 1 illustrates the fractionation of the products obtained with sodium tripolyphosphate. The

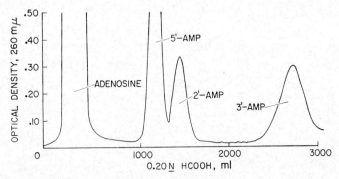

Fig. 1. Fractionation of the products of the phosphorylation of adenosine with $Na_5P_3O_{10}$ in water. See table 1 for the reaction conditions.

identities of the isomers were established by standardization of the column with authentic samples, and by thin layer chromatography. In addition, the three isomers were shown to be degraded to adenosine by *E. coli* alkaline phosphatase. The 2′ and 3′ isomers were further shown to be resistant to periodate oxidation, while the 5′ isomer was oxidized. Table 1 summarizes the results of several experiments. Pyrophosphate was ineffective under the conditions studied. Tripolyphosphate and Graham's salt, however, were effective phosphorylating agents. An interesting *p*H effect was observed with Graham's salt which appeared to be absent when phosphorylation was carried out with tripolyphosphate. Not only was there a large increase in the yield of the reaction, but the isomeric composition of the AMP synthesized changed in a striking manner. Under slightly acid, or neutral conditions, the products of the phosphorylation were roughly what might have been predicted. The more reactive 5′ primary hydroxyl group was more readily phosphorylated than the 2′ or 3′ positions. As the *p*H was increased, however, the 2′ and 3′ positions were phosphorylated almost to the exclusion of the 5′ hydroxyl group, a very surprising result. Table 2 compares the product compositions obtained in 0.5 N NaOH by the phosphorylation of adenosine and cytidine with Graham's salt. The isomeric distribution was constant, whether the nucleoside phosphorylated was adenosine or cytidine. This result would

TABLE 1

Compositions of reaction mixtures and products of the phosphorylation of adenosine

Phosphate	pH range[a]	Reflux (hr)	Relative yields of products			Total conversion of adenosine (%)
			5'-AMP	2'-AMP	3'-AMP	
$Na_4P_2O_7$	9.7–10.5	5.0	–	–	–	0
$Na_5P_3O_{10}$	7.5–8.0	5.0	54	18	28	1.1
$Na_5P_3O_{10}$	0.5 N NaOH	5.0	(60)	(27)	(13)	1.1[b]
Graham's salt	5.7–6.7	6.0	57	17	26	0.94
Graham's salt	9.7–10.5[c]	4.0	25	30	45	1.52
Graham's salt	0.5 N NaOH	5.0	6	40	54	10.4
Trimetaphosphate	5.4–6.4	5.0	–	–	–	<1
Trimetaphosphate	0.5 N NaOH	4.5	0	42	58	30.5

[a] The highest pH in each case was that at the start of the reaction; the lowest pH was measured at the conclusion of the reaction.

[b] Relative yields for this experiment are approximate.

[c] The reaction was run at 100°C in a sealed tube, in 0.75 N NH$_4$OH.

TABLE 2

Phosphorylation of adenosine and cytidine with Graham's salt
in 0.5 N NaOH

| Nucleoside | Relative yields of isomers | | | Total conversion[a] |
	5′	2′	3′	(%)
Adenosine	6	40	54	10.4
Cytidine	5	39	56	7.4

[a] Total conversions are not comparable, as there was a difference
in the scale of the two reactions.

tend to eliminate the possibility that some novel kind of interaction between the heterocyclic ring and the 5′ hydroxyl group might be hindering the phosphorylation of the primary position.

Recently, Feldmann reported the phosphorylation of several alcohols and carbohydrates by sodium trimetaphosphate in alkaline solution [5]. Since long chain polyphosphate preparations are known to contain small amounts of trimetaphosphate, and to produce trimetaphosphate upon hydrolysis in neutral or alkaline solution, the possibility existed that the observed change in the isomeric distribution of the products obtained in our experiments, might be due to the formation of trimetaphosphate in the reaction mixture [6]. To test this hypothesis, sodium trimetaphosphate was synthesized by annealing Graham's salt at 520°C for 12 hr. The product was found to be chromatographically homogeneous, and was resistant to the action of alkaline phosphatase, indicating the absence of end groups. Table 1 summarizes the data obtained with both linear and metaphosphates. It appears that the pattern of products obtained by the phosphorylation of adenosine with trimetaphosphate in 0.5 N NaOH is essentially the same as that produced under identical conditions with Graham's salt, although the yield is threefold higher with pure trimetaphosphate. These data suggest very strongly that it is indeed the formation of trimetaphosphate which accounts for both the increase in yield in basic solution, as well as the shift in product composition. The formation of trimetaphosphate rings under the conditions employed in these reactions has subsequently been substantiated by thin layer chromatography. The specificity of the phosphorylation appears to require the vicinal hydroxyl grouping of the 2′ and 3′ positions. Thus, when deoxyadenosine was reacted with sodium trimetaphosphate under conditions identical to those which produced a 30% yield of 2′ (3′)-AMP, the deoxynucleoside was only phosphorylated to the extent of 2%. The relative proportions of 5′-dAMP and 3′-dAMP were 37% and 63%, respectively. A similar specificity is suggested by Feldmann's report that the yield of phosphate ester, produced in alkaline solution with trimetaphosphate, decreases with separation of the hydroxyl groups in the series: ethanediol (1,2), propanediol (1.3), and butanediol (1.4). There have also been reports of specific phosphorylation of the 2′ and 3′ positions in nucleosides by phosphorus oxychloride in basic solution [7]. Since trimetaphosphate rings are possible products of the hydration of phosphorus oxychloride, this reaction may be a further example of the specificity of trimetaphosphate for vicinal hydroxyls (or at least for a *cis* *vic*-glycol) [8].

Any reagent which displays such selectivity in the synthesis of nucleotides is of considerable interest with regard to the origin of nucleic acids. Although it is highly unlikely that phosphorylation in 0.5 N NaOH is a process which occurred on the primitive earth, the mechanism of this reaction is extremely interesting, and is under further study. The most significant observation is that condensed phosphates, which are known to be potent phosphorylating agents under anhydrous conditions, can also phosphorylate nucleosides in aqueous solution. Such activity, over a wide range of pH, is particularly relevant to prebiological chemistry.

References

[1] F. Lipmann, in The origins of prebiological systems and of their molecular matrices, S. W. Fox, ed. (Academic Press, New York, 1965) p. 259.
[2] M. E. Jones and F. Lipmann, Proc. Natl. Acad. Sci. U.S. *46* (1960) 1194.
[3] J. R. Van Wazer, Phosphorus and its compounds, vol. I (Interscience, New York, 1958) p. 653.
[4] J. Aurenge, M. Degeorges and J. Normand, Bull. Soc. Chim. France (1964) 508.
[5] W. Feldmann, Chem. Ber. *100* (1967) 3850.
[6] E. Thilo, Angew. Chem. Intern. Ed. *4* (1965) 1961.
[7] G. R. Barker and G. E. Foll, J. Chem. Soc. (1957) 3798.
[8] M. Viscontini and G. Bonetti, Helv. Chim. Acta *34* (1951) 2435.

SYNTHESIS OF PYROPHOSPHATE UNDER
PRIMITIVE EARTH CONDITIONS*

STANLEY L. MILLER and MICHAEL PARRIS

Department of Chemistry, University of California, San Diego, La Jolla, California, U.S.A.

It is reasonable to assume that some of the polymerization and dehydration reactions of simple organic compounds on the primitive Earth were carried out by high-energy phosphates. Experiments based on this assumption have been performed by Schramm et al. [1], who reported that ethyl metaphosphate can effect a number of polymerizations including the synthesis of polynucleotides from nucleotides. These polymerizations are carried out in anhydrous dimethylformamide as solvent, and they apparently require closely controlled conditions since they are difficult to repeat in detail [2–4]. Ponnamperuma, Sagan and Mariner [5] have synthesized adenosine triphosphate from ADP, AMP or adenosine in dilute aqueous solutions using ethyl metaphosphate and ultra-violet light. The role of the ultra-violet light is not clear, since the ethyl metaphosphate is already a high-energy phosphate.

Ethyl metaphosphate is prepared from phosphorus pentoxide, diethyl ether and chloroform. It is most unlikely that such a polymer could have occurred on the primitive Earth, at least if synthesized in this way. The most unlikely aspect is not so much the diethyl ether or chloroform, but rather the P_2O_5. Phosphorus pentoxide does not occur naturally on the Earth, and if it were formed by some thermal process it could not be made available for synthetic reactions before it was hydrated by traces of moisture. The occurrence of polyphosphoric acid, which has been used in a 'thermal model of prebiological chemistry' [6–8], is also improbable for similar reasons.

It would be more plausible to assume the presence of meta-or poly-phosphates. These compounds are well known [9] and easily prepared by heating alkali metal or calcium dihydrogen phosphates above $\sim 300\,^{\circ}C$. However, metaphosphate minerals are not known to occur in nature. This is not surprising since $Ca(H_2PO_4)_2$, which is the starting material for a thermal dehydration to calcium metaphosphate, is itself not a known mineral [9]. Pyrophosphate could be formed by the thermal dehydration of brushite $(CaHPO_4 \cdot 2H_2O)$ or monetite $(CaHPO_4)$, which are relatively rare minerals,

* Reprinted by permission from Nature *204* (1964) 1248.

but this apparently does not occur since not a single pyrophosphate mineral is known [10–12].

Although pyro- or meta-phosphate minerals may eventually be discovered, it is clear that they do not occur on an extensive scale. Greater volcanic activity on the primitive Earth would not change this conclusion. The phosphate in igneous and metamorphic rocks, which would be the most likely places to find thermally produced metaphosphates, occurs principally as hydroxyl apatite $[Ca_{10}(PO_4)_6(OH)_2]$ with varying degrees of substitution of the hydroxide by fluoride and chloride. The principal phosphorus mineral in sedimentary rocks is also apatite.

The reducing conditions on the primitive Earth would not result in reduced phosphorus compounds to any great extent, since all the lower oxidation states of phosphorus are thermodynamically unstable in an aqueous environment [13]. Schreibersite (Fe_3P) occurs in the iron phase of meteorites, but the phosphorus would be oxidized to phosphate by water.

We have therefore looked for a low-temperature synthesis of pyrophosphate using various compounds with high free energies of hydrolysis that might have occurred on the primitive Earth, and we have found that pyrophosphate can be synthesized by the reaction of cyanate with hydroxyl apatite. Recently, Steinman, Lemmon and Calvin [14] have used cyanamide and its dimer to synthesize glucose-6-phosphate, pyrophosphate and AMP. These dehydration reactions were carried out in dilute solution at a pH of 2.

Experimental. The apatite was prepared by adding 0.04 M sodium dihydrogenphosphate to 0.025 M $CaCl_2$, both buffered at pH 8 with tris [tris(hydroxymethyl) aminomethane]. The apatite formed was centrifuged and washed several times with water. This apatite was finely divided, and its X-ray powder pattern showed good crystallinity. The apatite was suspended in 0.04 M tris buffer (0.09 M in phosphate if dissolved), and the pH was adjusted with hydrochloric acid or sodium hydroxide. The Ca^{2+} concentration in solution was brought to 0.01 M with calcium chloride, and the cyanate was added as KNCO. The mixtures were shaken in a water bath at 35°C for 20 days to complete the reaction. The apatite and calcium pyrophosphate formed were then dissolved with ethylenediamine tetraacetic acid, and this mixture was chromatographed on Dowex 1 [15]. The pyrophosphate in the fractions was hydrolysed, and the phosphate was determined by the Fiske–SubbaRow procedure [16]. The identification of product was confirmed by paper chromatography [17], and by comparing the infrared spectrum with an authentic sample of calcium pyrophosphate.

The yield of pyrophosphate as a function of pH with 0.001 M KNCO is

shown in fig. 1. The yields are expressed as moles of pyrophosphate synthesized per mole of cyanate added, and the maximum is 27 per cent at pH 5.6.
In addition to the pyrophosphate, a small amount of triphosphate (about 5
per cent of the pyrophosphate) was obtained. The yields of pyrophosphate
from concentrated phosphate buffers and cyanate are very low (fig. 1).

Although the synthesis of pyrophosphate is reproducible, there is considerable variation in the yields with the apatite reaction as would be expected
with a surface reaction. One preparation of apatite gave lower yields and did
not show the pH maximum.

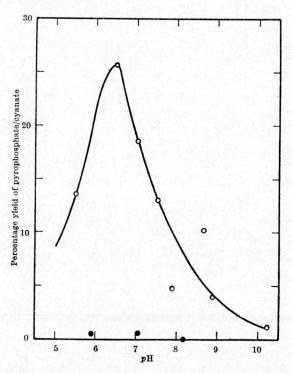

Fig. 1.   The yield of pyrophosphate as a function of pH. Apatite $+0.001$ M KNCO,
0.05 M tris buffer (0.1 M borate added for pH>8) ($\bigcirc$); 0.64 M sodium phosphate
buffer $+0.1$ M KNCO ($\bullet$). Samples shaken at 35 °C for 20 days.

The yields with apatite are not affected appreciably by the concentration
of Ca^{2+} or by adding Mg^{2+}. Pyrophosphate is also synthesized with apatite
prepared in the absence of tris and buffered with phosphate. The effect
of fluoride, trace elements and various organic compounds on this reaction
is being investigated.

The free energy of hydrolysis of cyanate to NH_4^+ and HCO_3^- is about -12 kcal at a pH of 8. There are a number of other compounds such as hydrogen cyanide, thiocyanate, cyanamide, and cyanogen which also have a sufficient free energy of hydrolysis to synthesize a pyrophosphate bond and which could have occurred on the primitive Earth. Under the conditions used, hydrogen cyanide, thiocyanate and cyanamide did not react with the apatite to give pyrophosphate. Cyanogen gave significant yields of pyrophosphate with apatite and with soluble phosphate, the yields being greater with the apatite. However, this reaction gives erratic results and is being investigated further.

Cyanate is a reasonable compound to have occurred in the oceans of the primitive Earth. Urea has been synthesized from methane, ammonia and water by electric discharges [18] and by high-energy electrons [19]. The urea in these experiments was probably synthesized from ammonia and cyanate by the Wöhler synthesis. This reaction is reversible [20], $NH_4^+ + NCO^- \rightleftarrows$ $CO(NH_2)_2$, $K_{25}^0 = 3 \times 10^4$. Although most of the cyanate produced by various sources of energy would react with the NH_4^+ in the oceans, the urea produced thereby would on decomposition yield a low steady-state concentration of cyanate to react with the apatite.

This synthesis could reasonably occur by the reaction of cyanate with phosphate on the surface of the apatite to form carbamyl phosphate:

$$HN{=}C{=}O + HPO_4 - 2 \rightleftarrows \underset{\displaystyle O \quad \quad O^-}{\overset{\displaystyle NH_2 \quad O}{C{-}O{-}P{-}O^-}}$$

Carbamyl phosphate ordinarily decomposes back to cyanate and orthophosphate. The equilibrium constant in solution is 3.1 [21]. The reaction of carbamyl phosphate with another phosphate at the $O-P$ bond rather than at the $C-O$ bond would give pyrophosphate by the displacement of carbamate,

$$\underset{\displaystyle O \quad \quad OH}{\overset{\displaystyle NH_2 \quad O}{C{-}O{-}P{-}O^-}} + \underset{\displaystyle OH}{\overset{\displaystyle O}{O^-{-}P{-}O^-}} \longrightarrow \underset{\displaystyle O}{\overset{\displaystyle NH_2}{C{-}O^-}} + \underset{\displaystyle OH \quad \quad OH}{\overset{\displaystyle O \quad \quad O}{O^-{-}P{-}O{-}P{-}O^-}}$$

The carbamate would then decompose to carbon dioxide and ammonia. These equations refer to reactions taking place on the surface, probably involving chelation by calcium, and the states of ionization have been written arbitrarily.

Jones and Lipmann [21] have proposed that carbamyl phosphate may have been an important source of high-energy phosphate bonds on the primitive Earth. The data here presented support this hypothesis, provided the reaction occurs on the surface of the most abundant phosphate mineral.

It has been pointed out by Gulick [22] and emphasized by Miller and Urey [13] that the synthesis of organic phosphates on the primitive Earth presents difficulties because of the very low solubility of apatite in the ocean $(3 \times 10^{-6}$ M phosphate). The results reported here suggest that a solubilization mechanism for calcium phosphates may not have been needed. The phosphate could have been made available for primitive syntheses by reactions taking place on the surface of phosphate minerals, the concentration mechanism for this relatively non-abundant element being a simple precipitation.

It would appear that pyrophosphate minerals, which are not found on the Earth at the present time, may have occurred in significant amounts on the primitive Earth during the period when the organic compounds leading to the origin of life were being synthesized. This pyrophosphate may have played an important role in polymerization and dehydration reactions of these organic compounds.

This work was supported by U.S. National Science Foundation grant G–22000. We thank Prof. G. Arrhenius for obtaining the X-ray powder patterns.

References

[1] G. Schramm, H. Grötsch and W. Pollman, Angew. Chem. 74 (1962) 52; Intern. Ed. 1 (1962) 1.
[2] J. A. Carbon, Chem. Ind. (1963) 529.
[3] N. K. Kochetkov, E. I. Budowsky, V. D. Domkin and N. N. Khromov-Borissov, Biochim. Biophys. Acta 80 (1964) 145.
[4] B. P. Gottich and O. I. Slutsky, Biochim. Biophys. Acta 87 (1964) 163.
[5] C. Ponnamperuma, C. Sagan and R. Mariner, Nature 199 (1963) 222.
[6] S. W. Fox, Science 132 (1960) 200.
[7] S. W. Fox and K. Harada, Science 133 (1961) 1923.
[8] A. Schwartz and S. W. Fox, Biochim. Biophys. Acta 87 (1964) 694.
[9] J. R. Van Wazer, Phosphorus and its compounds, Vol. 1 (Interscience, New York, 1958).
[10] G. Palache, H. Berman and C. Frondel, Dana's system of mineralogy, seventh ed., Vol. 2 (John Wiley and Sons, New York, 1951).
[11] V. M. Goldschmidt, Geochemistry (Oxford Univ. Press, 1954) p. 454.
[12] K. Rankama and T. G. Sahama, Geochemistry (Univ. Chicago Press, 1950), p. 584.

[13] S. L. Miller and H. C. Urey, Science *130* (1959) 245.
[14] G. Steinman, R. M. Lemmon and M. Calvin, Proc. Natl. Acad. Sci. U.S. *52* (1964) 27.
[15] S. Lindenbaum, T. V. Peters, jr and W. Rieman, Anal. Chim. Acta *11* (1954) 530.
[16] C. H. Fiske and Y. SubbaRow, J. Biol. Chem. *66* (1925) 375.
[17] E. Karl-Kroupa, Anal. Chem. *28* (1956) 1091.
[18] S. L. Miller, Biochim. Biophys. Acta *23* (1957) 480.
[19] C. Palm and M. Calvin, J. Amer. Chem. Soc. *84* (1962) 2115.
[20] I. A. Kemp and G. Kohnstam, J. Chem. Soc. (1956) 900.
[21] M. E. Jones and F. Lipmann, Proc. Natl. Acad. Sci. U.S. *46* (1960) 1194.
[22] A. Gulick, Am. Scientist *43* (1955) 479; Ann. N.Y. Acad. Sci. *69* (1957) 309.

POLYNUCLEOTIDE REPLICATION AND THE
ORIGIN OF LIFE

L. E. ORGEL and J. E. SULSTON

The Salk Institute of Biological Sciences, San Diego, Calif. 92112, U.S.A.

The replication of polynucleotides in living organisms appears to depend upon the formation of the specific hydrogen-bonded base pairs adenine:thymine (or uracil) and guanine:cytosine. The existence of these base pairs was first postulated by Crick and Watson, and has since been amply confirmed in a variety of systems (for review see [1]). Most significantly, specific pairing of monomers has been demonstrated. Thus, if suitably protected derivatives of the four bases are dissolved in an organic solvent, the strongest interactions are those between guanine and cytosine on the one hand and between adenine and uracil on the other [2–5]. The protected guanine–cytosine pair has been crystallized [6].

Purine nucleosides and nucleotides form complexes with the complementary homopolymers under suitable conditions [7–9]. For example, A* and poly U yield the triple helix A:2 poly U, which is very similar to the triple helix poly A:2 poly U. The monomer–polymer complex has a fairly sharp melting point, indicating cooperative interaction, but is less stable than the polymer complex. No complex between a pyrimidine monomer and a purine polymer has been found, but oligouridylates of sufficient length form stable helices with poly A [1].

The existence of such complexes suggests that polymers might be used as templates in the non-enzymic condensation of complementary monomers or oligomers. Naylor and Gilham found that the self-condensation of hexathymidylate in the presence of a water soluble carbodiimide to yield dodecathymidylate was enhanced by poly dA [10]. In this laboratory we have studied the template directed condensation of mononucleotides, and have examined the specificity of the reaction.

Summary of results. Under conditions in which the triple helices A:2 poly U and pA:2 poly U are known to be stable, poly U facilitates the condensation

* Abbreviations: A, adenosine; G, guanosine; C, cytidine; U, uridine; poly U, polyuridylate; dA, deoxyadenosine; pA, adenosine-5'-phosphate; ApA, adenylyladenosine; pppA, adenosine-5'-triphosphate; ImpA, adenosine-5'-phosphorimidazolide; ImpU, uridine-5'-phosphorimidazolide.

of A with pA in the presence of a water-soluble carbodiimide [11]. The yield of ApA is about ten times greater than that in the absence of poly U; two thirds of the ApA is $(2' \rightarrow 5')$ linked, one third is $(5' \rightarrow 5')$ linked and only 2% is the natural $(3' \rightarrow 5')$ isomer. Under the same conditions, poly U has no effect on the condensation of G, C or U with pA, nor does poly C have any effect on the condensation of A with pA [12]. Poly U also facilitates the self-condensation of pA [11], and the condensation of pA with oligomers such as ApApApA [13]. Thus both chain initiation and extension can occur.

Similarly, poly C facilitates the self-condensation of pG and the condensation of pG with G. Poly C does not affect the reaction of pG with A, C or U. In each case, most of the phosphodiester bonds formed are $(5' \rightarrow 5')$, and therefore reaction under these conditions could not yield polymers efficiently. However, the results serve to demonstrate that the Watson–Crick pairing rule is obeyed.

Poly U catalyzes the self-condensation of pdA and the condensation of dA with pdA [15]. The products are mainly $(5' \rightarrow 5')$ linked, although the proportion of $(3' \rightarrow 5')$ isomers is higher than in the ribo series.

In order to establish a plausible system for prebiotic replication, it is necessary to replace the carbodiimide by a condensing agent likely to have been present on the primitive earth [16]. Cyanamide and cyanoguanidine cannot easily be tested, since they react so slowly at the low temperatures required for the formation of stable helices. Cyanogen is sufficiently reactive at $0°C$, but does not bring about template-directed synthesis.

Alternatively, a preformed activated nucleotide might be used. pppA forms a stable complex with poly U, but then hydrolyzes without forming appreciable amounts of oligonucleotides. Adenosine–5′–phosphorimidazolide (ImpA), however, reacts very efficiently to give phosphodiester bonds. Thus, ImpA condenses with A in the presence of poly U to yield 50% (based on ImpA) of ApA [17], in the experiments described above, the corresponding efficiency based on carbodiimide is about 2%. Furthermore, 96% of the ApA formed from ImpA is $(2' \rightarrow 5')$ linked and only 2% $(5' \rightarrow 5')$ linked. Since $(5' \rightarrow 5')$ phosphodiester bonds cannot be used in chain propagation, their suppression would be of considerable advantage. ImpA also reacts efficiently with pA [17] and with ApApApA [13].

When DL-adenosine reacts with ImpA on a poly U template, D-adenosine is incorporated ten times more rapidly than L-adenosine [18]. This observation provides a basis for the segregation of D- and L-nucleotides at an early stage in biochemical evolution.

Poly A does not facilitate the condensation of ImpU with U or UpUpU.

This result is in accord with the failure to find stable complexes between U and poly A. The difficulty may perhaps be overcome by the replication of alternating purine-pyrimidine sequences, or by the incorporation of pre-formed oligomers: the latter has already been demonstrated by Naylor and Gilham [10].

The template reactions described above lead predominantly to $(2' \rightarrow 5')$ phosphodiester links rather than the naturally occurring $(3' \rightarrow 5')$ links. However, it is known that oligo-$(2' \rightarrow 5')$-adenylates form helices with poly U, and that these complexes are only slightly less stable than those formed by oligo-$(3' \rightarrow 5')$-adenylates [9]. It is therefore possible that primitive poly-nucleotides contained both $(3' \rightarrow 5')$ and $(2' \rightarrow 5')$ links; the selection of a $(3' \rightarrow 5')$ linked ribose phosphate backbone may have awaited the evolution of enzymes. The slightly greater stability of the complexes or enhanced resistance to hydrolysis are possible explanations for this selection.

We are continuing to examine possible prebiotic activating agents, and to define the conditions under which pyrimidines can be incorporated. It must be pointed out that template synthesis requires nucleotide concentrations of a few hundredths molar, rather than the very dilute solutions available on the primitive earth. Perhaps both template and monomers were adsorbed onto a mineral or an organic surface, or perhaps replication was restricted to regions where concentration or freezing had taken place.

Discussion. The most difficult task facing any theory of the origins of life is to explain the evolution of the genetic code. It is widely believed that the genetic apparatus of the first organisms was not made up of proteins alone, since it seems unlikely that proteins can replicate. While nucleic acid replication in the absence of enzymes now seems much more plausible, nucleic acids are not known to act as catalysts. Proteins function but apparently cannot repli-cate; nucleic acids replicate but apparently cannot function. Thus we seem forced to postulate a connection between polynucleotides and polypeptides from the beginning.

This approach leads to new difficulties so severe that it has never been carried very far. Protein synthesis involves a complex system of soluble enzymes and tRNA's in addition to a highly evolved nucleoprotein particle, the ribosome. Clearly a primitive protein synthesizing apparatus could have been less complicated, but it is hard to see how any reliable system capable of synthesizing polypeptides of defined sequence and length could be simple.

Further progress is made possible only by ascribing to polynucleotides or polypeptides properties which have not been demonstrated experimen-tally, and which usually seem implausible. For example, if there is a repeating peptide xyxyxy... which facilitates nucleotide replication, and if a repeating polynucleotide ABABAB... was formed spontaneously on the primitive earth then an explanation of the evolution of a simple genetic code which assigned A to xyx and B to yxy might be possible.

Here we proceed quite differently. The evolution of the protein-synthesizing system would seem less incredible if its components could have had a function independent of protein synthesis but nonetheless helpful to polynucleotide replication. It would be even better if the postulated function could have evolved *by small steps* and led ultimately to the genetic code. The attempt to define such a function and to discover evidence for it is at least an interesting intellectual excercise.

Perhaps there are some clues in contemporary biochemistry. Many writers have been struck by the important role which non-genetic polynucleotides play in protein synthesis. The ribosome contains more than 50% of RNA and yet performs the sort of task which we have come to associate with systems of proteins. Indeed, the ribosome is clearly assisted by a large number of proteins whose spatial relations it seems to organize. Does this indicate that the ribosome once functioned with very few proteins, or perhaps none at all? It seems unlikely that the protein of the ribosome is more primitive than the RNA, and at least possible that the opposite is true.

The widespread use of nucleotide coenzymes in metabolic pathways is puzzling. It is clear, for example, that N-methyl nicotinamide would function perfectly well in dehydrogenation reactions. Of course, the adenylic acid portion is now employed as a handle to recognize the molecule, but it is a rather unexpected handle. The nucleotide coenzymes look as though they are meant to attach to polynucleotides by base-pairing rather than to enzymes.

This leads us to ask whether the ribosomes and the coenzymes are biochemical fossils. Is it possible that the first organisms consisted of polynucleotides which replicated and carried out catalytic functions with the assistance of small molecules which they scavenged from their environment? Would glycolysis be possible catalyzed by NAD, polynucleotides and perhaps other small molecules? Could polysaccharides be constructed from UDP-glucose, etc.? Clearly we cannot answer these questions, but since we know that RNA molecules can adopt a complicated, three-dimensional structure, an affirmative answer does not seem impossible.

Experiments described by Miller and others [19, 20], suggest that amino acids such as glycine, serine and aspartic acid were formed on the primitive earth. If our guess is correct, these molecules too would have been used by the polynucleotide system. The attachment of amino acids to appropriate polynucleotides could have conferred a selective advantage on the latter even in the absence of protein synthesis.

Once granted the presence of self-replicating polynucleotides specifically attached to amino acids the major obstacle to the evolution of the genetic code is overcome. The genetic code could have evolved bit-by-bit. Many different evolutionary sequences can be envisaged. Here we outline the earliest stage of one such sequence to illustrate the principle. The details are not significant.

Suppose the original system included two families of polynucleotides which bound and utilized serine and glutamic acid, respectively, and that the dipeptide serylglutamic acid could be utilized by the system better than the separate amino acids. Then if, by mutation, the two polynucleotides "learned" to associate together to give the dipeptide, "organisms" containing both would be at an advantage over their neighbours and outgrow them. If by further mutation these RNA's "learned" to base-pair with another stretch of RNA to make the original association firmer, the development of the "messenger-tRNA" system would have been initiated.

These fantasies are based on the proposition that polynucleotide molecules are capable of binding specifically certain small molecules to give complexes that are catalytically active. This proposition and the related one that certain polynucleotide sequences are themselves catalytically active have the merit that they can be tested experimentally. We are undertaking such experiments.

References

[1] G. Felsenfeld and H. T. Miles, Ann. Rev. Biochem. *36* (1967) 407.
[2] R. H. Hamlin, R. C. Lord and A. Rich, Science *148* (1965) 1734.
[3] Y. Kyogoku, R. C. Lord and A. Rich, Science *154* (1966) 518.
[4] R. R. Shoup, H. T. Miles and E. D. Becker, Biochem. Biophys. Res. Commun. *23* (1966) 194.
[5] L. Katz and S. Penman, J. Mol. Biol. *15* (1966) 220.
[6] H. M. Sobell, K. Tomita and A. Rich, Proc. Natl. Acad. Sci. U. S. *49* (1963) 885.
[7] F. B. Howard, J. Frazier, M. F. Singer and H. T. Miles, J. Mol. Biol. *16* (1966) 415.
[8] W. M. Huang and P. O. P. Ts'o, J. Mol. Biol. *16* (1966) 523.
[9] A. M. Michelson and C. Monny, Biochim. Biophys. Acta *109* (1967) *149.*
[10] R. Naylor and P. T. Gilham, Biochemistry *5* (1966) 2722.
[11] J. Sulston, R. Lohrmann, L. E. Orgel and H. T. Miles. Proc. Natl. Acad. Sci. U. S. *59* (1968) 726.
[12] J. Sulston, R. Lohrmann, L. E. Orgel and H. T. Miles, Proc. Natl. Acad. Sci. U. S. *60* (1968) 409.
[13] H. Schneider-Bernloehr, unpublished experiments.
[14] J. Sulston, R. Lohrmann, L. E. Orgel, H. Schneider-Bernloehr, B. J. Weimann and H. T. Miles, in preparation.

[15] H. Schneider-Bernloehr, R. Lohrmann, J. Sulston, B. J. Weimann, L. E. Orgel and H. T. Miles, J. Mol. Biol. *37* (1968) 151.
[16] R. Lohrmann and L. E. Orgel, Science *160* (1968) 64.
[17] B. J. Weimann, R. Lohrmann, L. E. Orgel, H. Schneider-Bernloehr and J. Sulston, in preparation.
[18] H. Schneider-Bernloehr, R. Lohrmann, L. E. Orgel, J. Sulston and B. J. Weimann, in preparation.
[19] S. L. Miller, J. Am. Chem. Soc. *77* (1955) 2351.
[20] S. W. Fox, ed., The origins of prebiological systems and of their molecular matrices (Academic Press, New York, 1965).

SYNTHESIS AND PROPERTIES OF
POLYARABINONUCLEOTIDES

G. SCHRAMM[†]

Max-Planck-Institut für Virusforschung, Tübingen, Germany

The cell synthesizes larger molecules mostly by elimination of phosphates or phosphate esters from two components. This is schematically shown by the following formulae:

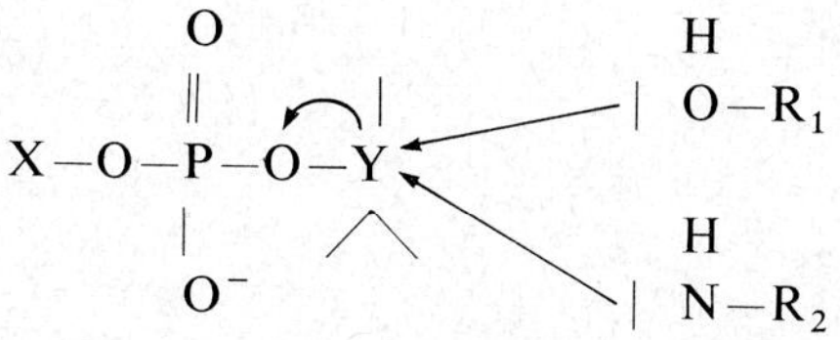

$$X-O-P(=O)(O^-)-O-Y \quad \begin{cases} H | O-R_1 \\ H | N-R_2 \end{cases}$$

X can be H, phosphate, pyrophosphate, nucleoside or nucleoside monophosphate. Y can be a C-atom (synthesis of polysaccharides, aminoacyl-t-RNA, nucleosides and amides), or a P-atom (synthesis of polynucleotides). On phosphorylation of the hydroxy group of one component, electrons are withdrawn from the central atom which can now be attacked by a nucleophilic group of the other component, for example an OH, NH, SH group or a carbanion. The electron withdrawal might be enhanced by metal ions and enzymes. In nonenzymatic syntheses, the central atom can be made more electrophilic (positive) by suitable substitution of the phosphate group.

Some years ago we found that the condensation occurring in the cell could be successfully simulated by use of polyphosphate esters [1]. Originally we used ethylpolyphosphates, but we observed that this agent causes ethylation in addition to phosphorylation of the amino and hydroxy groups. This undesirable side reaction can be avoided by the use of phenylpolyphosphates, in which the O − C bond is never cleaved and which phosphorylate only.

Initially the phosphate esters were prepared by the reaction of dialkylethers with phosphorus pentoxide according to Langheld [2], but this procedure is difficult and cannot be applied to arylethers. It has been found that phenylpolyphosphates can easily be prepared by fusion of triphenylphosphates with phosphorus pentoxide at 200–300 °C [3]

[†] Deceased, see dedication before Preface.

$$P_4O_{10} + 2 \ (PhO)_3PO \rightarrow 6 \ \left(-\overset{\displaystyle O-Ph}{\underset{\displaystyle O}{\overset{|}{\underset{\|}{P}}}} - \right).$$

The resulting mixture of polyphenylphosphates can be analyzed by nuclear magnetic resonance of the phosphorus. The chemical shifts of the resonance signal are different for orthophosphate esters (P_o), end groups (P_e), middle groups (P_m) and branched groups (P_b). Therefore polyphenylphosphate can be characterized by the composition of these units.

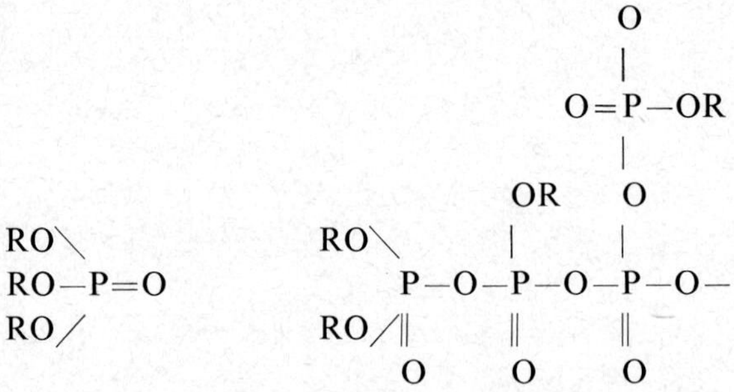

$$
\begin{array}{cccc}
 & & & O \\
 & & & | \\
 & & & O = P - OR \\
 & & & | \\
 & & OR & O \\
RO\diagdown & RO\diagdown & | & | \\
RO - P = O & P - O - P - O - P - O - \\
RO\diagup & RO\diagup\| & \| & \| \\
 & O & O & O \\
P_o & P_e & P_m & P_b
\end{array}
$$

More recently we found a very simple method for the preparation of phenyl-polyphosphates by heating of orthophosphoric acid with phenol. The products are less well defined than the compounds described above, but suitable for condensing reactions. If phenylpolyphosphate is exposed to humidity or other nucleophilic agents the signal of the branched phosphorus P_b disappears very rapidly. This is the most reactive form. After consumption of P_b the phosphorylation proceeds more slowly until all P—O—P bonds are cleaved by the nucleophilic agent. The reaction rate can be determined by titration of the P—OH groups in nonaqueous solvents. Comparison with the reaction rate of diphenylphosphochloridate $(Ph)_2POCl$ and other phosphorylating agents revealed that the branched form of the phenylphosphate is the most powerful agent. This is in agreement with the theory of Van Wazer [4].

We have studied the syntheses of many biologically interesting molecules with polyphosphate esters. Amino acids and oligopeptides can be combined to form polypeptides, monosaccharides can be polymerized to polysaccharides, monosaccharides can be condensed with purines and pyrimidines to nucleosides, and mononucleotides and oligonucleotides can be polymerized to polynucleotides.

In some cases the similarity to the biosynthesis is obvious. The nucleoside phosphorylase of bacteria catalyzes the formation of deoxynucleosides from deoxyribosyl-1-phosphate and purines by elimination of the phospate group.

This reaction can be simulated with phenylpolyphosphate. When deoxyribose or ribose are heated with adenine in presence of phenylpolyphosphate for a few minutes the 9-N-furanosyl-derivatives are formed in a smooth reaction yielding α- and β-anomer in nearly equal amounts [5]. Starting with acylated ribose we obtained only the β-anomer [6]. The mechanism of the reaction could be elucidated. At first the hydroxy group at the 1-C of the sugar is phosphorylated, then this phosphate ester reacts with the 9-N of the adenine and the substituted phosphate is eliminated. In the classical nucleoside synthesis of Fischer and Helferich [7] a halogenide at 1-C is the leaving group. In the classical method of nucleoside synthesis many steps are required for the introduction and removal of protecting groups, whereas the synthesis with polyphosphates can be carried out with the free sugars and the free bases. Instead of ribose or deoxyribose, arabinose [8], xylose [9] or glucose [9] can be used. The reaction also proceeds with uracil instead of adenine [9]. The reaction of ribose and deoxyribose with adenine demonstrates that under very simple conditions predominantly those nucleosides are formed which have the same configuration (except the anomery) as the natural building blocks of the nucleic acids. The ease of this reaction and the resemblance to the biochemical procedure have led us to the idea that polyphosphate esters were involved in the formation of nucleic acids in the early history of the earth [10]. This seems plausible because phosphates had to be present as constituents of the nucleic acids.

To support this hypothesis we studied the condensation of deoxynucleotides [11] and ribonucleotides [12] with polyphosphate esters. Schramm and Ulmer [12] described the condensation of uridylic acid with polyphenyl phosphate. By intermediary formation of poly-O^2-2'-cyclouridylates and inversion of the 2'-OH group polyarabinonucleotides are obtained. Because of the trans-position of the 2' and 3' OH groups 2', 3'-cyclic phosphates cannot be formed from polyarabinonucleotides (poly-ara-U). Therefore poly-ara-U shows a high stability in alkaline and acidic solution and is resistant to ribonuclease. This stability permits the removal of polyuridylic acid (poly-U) by hydrolysis. Degradation with snake venom diesterase and bacterial monophosphatase leads to 9-β-D-arabinosyl uracil (spongouridine) naturally occurring in sponges. From this enzymatic degradation and chemical experiments it is concluded that polyspongouridylic acids have 3'-5' diester linkages like the natural polynucleotides.

The chain lengths of our preparations were determined in different ways. Analysis of the various peaks in the DEAE cellulose chromatography yielded spongouridylates up to the pentanucleotide. Since the amounts of higher spongouridylates are small and separation is difficult the average chain length of this material was determined on a standardized column of Sephadex G-75 according to Hohn and Schaller [13]. The maximum of the distribution curve is near $n \sim 7$, but spongouridylates with $n = 10$–20 are also present. The average chain lengths found were monitored by molecular weight determinations in the ultracentrifuge. For the top fraction comprising chains with $n = 10$–20 an average molecular weight of 5100 was found corresponding to $n = 17$.

One should expect that polyarabinonucleotides have an intermediate position between deoxyribonucleotides and ribonucleotides. The resistance of their phosphate linkage to alkali and ribonuclease resembles that of DNA, however, in acidic solution the stability of the N-glucosidic linkage in the polyarabinonucleotides resembles that of RNA rather than that of DNA, because of the inductive influence of the 2'-OH. Therefore it is interesting to compare the physico-chemical properties of poly-ara-U with the corresponding polyribonucleotides and polydeoxyribonucleotides. The 2'-OH group has a large influence on the conformation and the helix-coil transition of polynucleotides. In general the helical structure is more stable and shows a higher melting temperature in single-stranded ribopolynucleotides than in single-stranded DNA. According to Ts'o [14] the higher stability can be explained by the hypothesis that the 2'-OH group hydrogen bonds with the 2-keto group of the pyrimidine and with the N-3 nitrogen of adenine. In poly-ara-U with $n = 7$ no hypochromicity was detected at all at room temperature and no hybridization could be observed after mixing with polyriboadenylic acid. pK measurements gave no indication for a hydrogen bond between 2'-OH and O^2 of the pyrimidine. The obvious lack of base stacking and hybridization with poly-A is probably not caused by the lack of the hydrogen-bond, but may be due to restriction of the freedom of revolution by the 2'-OH group pointing to the same side of the furan ring as the pyrimidine residue (in arabino-uridine both residues are in the β-position). The interactions of arabinonucleosides and polyarabinonucleotides with natural polynucleotides are under investigation.

A similarity of poly-ara-U to the natural polynucleotides is also indicated by interaction with the enzymes specific for DNA and RNA. Poly-ara-U is not degraded by pancreatic deoxyribonuclease. However, a weak competitive inhibition of this endonuclease is observed. In the presence of a 100 fold excess of poly-ara-U $(n = 7-10)$ the rate of degradation of DNA is decreased by 50%. A high susceptibility has been found to pancreatic ribonuclease. At the ratio poly-ara-U $(n = 7-10)$ to RNA $= 1 : 5$ the inhibition was 50%. Therefore the susceptibility to ribonuclease seems to be 5 times higher than that of the high molecular weight RNA. The inhibition depends on the chain length. Smaller oligonucleotides have a much weaker effect.

After dephosphorylation of the terminal phosphates poly-ara-U is bound to ribosomes and stimulates the adaptation of Phe-t-RNA. From the preliminary experiments [12] it cannot be decided whether poly-ara-U has the same or a weaker stimulation effect than poly-U of the corresponding chain length.

Poly-ara-U shows a weak priming activity with polynucleotide phosphorylase [15], but it is not certain that this is due to the arabino moiety of the molecule, since the preparation contained small amounts of ribouridylic acid on the 5'-terminal position. Poly-ara-U is completely stable against degradation by polynucleotide phosphorylase and does not inhibit the degradation of poly-U by this enzyme.

DNA-dependent bacterial RNA polymerase can accept RNA instead of DNA as template. Poly-ara-U inhibits the template activity of either poly-T or Poly-U, probably by displacement of the natural templates. It has been found that the arabino nucleoside triphosphates ara-CTP and ara-ATP competitively inhibit the polymerization of the ribonucleoside triphosphates [16]. Thus the inhibition at the substrate level resembles that of the templates.

Poly-ara-U is degraded neither by ribonuclease nor by polynucleotide phosphorylase. As far as known snake venom diesterase is the only enzyme which cleaves poly-ara-U. Since this enzyme does not occur in animal cells, a large life-time can be expected for poly-ara-U in vivo. By competition to the natural polynucleotides an effect can be expected on the nucleic acid metabolism. In vivo experiments [15] were done with chicken fibroblasts and mouse L-cells. When mouse L-cells are kept in a minimal medium for 5 hr the cells change from growing to resting state and the synthesis of DNA, RNA and protein and the activity of uridine kinase are diminished. If poly-ara-U is present during the incubation the biological activities are maintained at about the original level. This maintenance effect is probably not due to the induction of a protein synthesis, since it can also be observed in the presence of puromycin and cycloheximide. The uptake and incorporation of uridine into RNA is regulated by the uridine kinase [17]. Therefore the influence of poly-ara-U on the RNA metabolism might be explained by the interaction with uridine kinase. Poly-ara-U may be a positive effector for the kinase or an inhibitor for a specific phosphatase which counteracts the kinase. The direct or indirect activation could lead to a preservation of the RNA metabolism which supports the DNA and protein metabolism. Because of the numerous feed-back mechanisms involved in the polynucleotide metabolism the action of poly-ara-U on the transition from the resting to the growing state is open to further questions.

Studies on poly-ara-nucleotides may be helpful in achieving an understanding of the physical, chemical and biological properties of the natural polynucleotides. They may also give a hint to the abiogenic origins of nucleic acids. The stability of polyarabinonucleotides and their smooth formation from mononucleotides and polyphosphates suggest that polynucleotides of this type might have been the first matrices which induced the formation of further polynucleotides from monomers to oligomers. The intermediate position of the polyarabinonucleotides between RNA and DNA suggests further that they might have been a starting point for the evolution of both types of nucleic acids.

References

[1] G. Schramm, H. Grötsch and W. Pollmann, Angew. Chem. 74 (1962) 53.
[2] K. Langheld, Chem. Ber. 43 (1910) 1857; and 44 (1911) 2076.
[3] G. Schramm and H. Berger, Z. Naturforsch. 22b (1967) 587.
[4] J. R. van Wazer, Phosphorus and its compounds (Interscience, New York, 1958).
[5] G. Schramm, G. Lünzmann and F. Bechmann, Biochim. Biophys. Acta 145 (1967) 221.
[6] H. Köster and G. Schramm, unpublished.
[7] E. Fischer and B. Helferich, Chem. Ber. 47 (1914) 210.

[8] G. Schramm and G. Lünzmann, in press.
[9] Unpublished.
[10] G. Schramm, in The origins of prebiological systems and of their molecular matrices,
      S. W. Fox, ed. (Academic Press, New York, 1965) p. 299.
[11] Th. Hohn, Dissertation, Tübingen (1966).
[12] G. Schramm and I. Ulmer-Schürnbrand, Biochim. Biophys. Acta 145 (1967) 7.
[13] Th. Hohn and H. Schaller, Biochim. Biophys. Acta 138 (1967) 466.
[14] P. O. P. Ts'o, S. A. Rapaport and F. J. Bollum, Biochemistry 5 (1966) 4153.
[15] G. Schramm, I. Ulmer-Schürnbrand and Ch. Ullrich, in press.
[16] S. S. Cohen, Progr. Nucleic Acid Res. 5 (1966) 72.
[17] P. Hausen and H. Stein, European J. Biochem. 4 (1968) 401.

CARL A. RUDISILL LIBRARY
LENOIR RHYNE COLLEGE

POLYMERIZATION OF NUCLEOTIDES VIA DISPLACEMENT ON CARBON; ITS PREPARATIVE AND PREBIOTIC SIGNIFICANCE

JOSEPH NAGYVARY and ROBERTO PROVENZALE

Department of Biochemistry and Biophysics, Texas A and M University, College Station, Texas, U.S.A.

Unlike the stepwise synthesis of oligonucleotides, the progress in the chemical polymerization of nucleotides has been rather slow in spite of the considerable effort invested [1]. So far the major approach of phosphorylation [2] has been based on the activation of the phosphate by a great variety of activating agents followed by the esterification of a free alcoholic group (eq. 1).

$$ROH + X-\underset{\underset{OR'}{|}}{\overset{\overset{O}{\|}}{P}}-X \rightarrow RO-\underset{\underset{X}{|}}{\overset{\overset{O}{\|}}{P}}-OR' + X^- \qquad (1)$$

A drawback of this method as applied to polymerization is that the main reaction seems to be slowed down by the fixation of the growing chains by fully esterified pyrophosphate linkages, making the alcoholic group unavailable for the activated monoester. This difficulty has been eliminated by the utilization of intermediate triesters [3], however, the contribution of this idea to polymerization still remains to be seen. Several secondary reactions have also been observed, such as the formation of cyclic oligonucleotides and a few tertiary ester linkages and the reaction of the alcoholic groups with the activating agents.

A second approach of internucleotidic bond formation, a displacement reaction on an activated carbon (eq. 2), was attempted by Elmore and Todd [4], and Khorana [5], respectively, but it was not further developed after initial difficulties were encountered.

$$RO-\underset{\underset{O^-}{|}}{\overset{\overset{O}{\|}}{P}}-O^- + R'-X \rightarrow RO-\underset{\underset{O^-}{|}}{\overset{\overset{O}{\|}}{P}}-OR' + X^- \qquad (2)$$

In principle, when both the leaving group X and the phosphate are contained in the same molecule, a polymerization reaction can take place upon providing the necessary activation energy. In this sense, the most suitable self-condensing system should satisfy the following requirements: the starting material should be stable enough to allow isolation in highest purity, and the group X should possess sufficient reactivity as a leaving group under reasonable thermal activation. The advantages of such an approach are obvious: there is no need for added activating agents; also, very high concentrations may be used, or the solvent may be entirely omitted.

A particular case of carbon activation can be found in cyclonucleosides [6], which were recently introduced in several laboratories for the synthesis of the internucleotide linkage [7]. The following cyclonucleosides have already been utilized:

No polymerization based on this so called "cyclonucleoside method" * has been reported so far. Here the obstacle is the preparation and purification of the suitably phosphorylated cyclonucleosides, or by another name, cyclonucleotides.

Here we describe the polymerization experiments which we carried out on O^2, 5'-cyclouridine-2', 3'-cyclic phosphate. This compound can be prepared from 5'-O-mesyl uridine-2', 3'-cyclic phosphate [8] and 3 equivalents of N,N'-dicyclohexyl-4-morpholino-carboxamidine when reacted in dry dimethylformamide for two days at room temperature, or overnight at 40 °C. It was purified by chromatography on silica gel in the solvent system n-butanol-acetone-water 4:2:1(v/v/v). The purification was facilitated by the resistance of O^2, 5'-cyclouridine-2', 3'-cyclic phosphate to ribonuclease. The ultimate proof of purity of our material was provided by its successful crystallization from dimethylformamide–tri-n-butylamine in the form of colorless needles, which exhibited UV, ORD, and NMR spectra consistent with this system.

Two kinds of polymerization experiments were carried out: in solution and in solid state. The latter is the first example of a solid state polymerization of a nucleotide.** The best conditions were found as follows. A 0.3 M dimethylformamide solution of carboxamidinium or tri-n-butylammonium cyclouridylate containing one half equivalent of tri-n-butylamine was kept in a sealed tube at 80 °C for three days and then the temperature was raised to 100 °C within three days and kept at this temperature for two more days. The reaction product was analyzed for chain length by gel filtration on Sephadex G-75 according to Hohn and Schaller [9]. A broad distribution of chain length ranging from 10 to about 50 nucleotide units was found, with an average of about 18 (fig. 1, curve A). The solid state polymerization was performed on the crystalline carboxamidinium and the amorphous lithium salts beginning at 100 °C and then raising the temperature to 140 °C over the period of two days. The chain length distribution of the product seems to depend upon the nature of the cation (fig. 1, curves B and C).

The structural work on these polymers was facilitated by our previous studies using dinucleoside phosphates as models. Scheme 1 illustrates the probable mechanism of the complex process of condensing the O^2, 5'-cyclouridylate to a polymer containing the isomeric O^2, 2'-cyclouridine moieties, which are linked together with $3' \rightarrow 5'$ phosphodiester linkages.

* Also named 'anhydronucleoside method' [7].
** This solid state synthesis is not to be confused with syntheses on solid state support in solution.

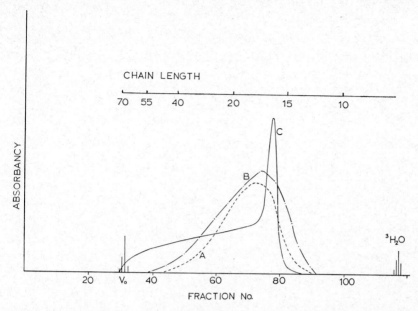

Fig. 1. Distribution of the polymerization products from O^2, 5'-cyclouridine-2', 3'-cyclic phosphate by a modified method of Hohn and Schaller [9] on a Sephadex G-75 column (106×1.6 cm) in 1 M triethylammonium bicarbonate, pH 8.2; 3 ml fractions. Curves A, B and C depict experiments from the carboxamidinium salt in solution, the same salt in the solid state and the lithium salt in the solid state, respectively.

The first step in the reaction produces the very labile 2', 3'-cyclic 5' triester (scheme 1, upper part of formula II) which rearranges as soon as the C-2 oxygen becomes available. The great susceptibility of this cyclic triester to hydrolysis is responsible for the presence of a varying number of ribomoieties in the primary product. The ribouridine-2' (3') phosphodiester linkages contained in this heteropolymer, together with the cyclonucleoside ether linkages, are then hydrolyzed by alkali, giving a homopolymer of arabinouridylates with one ribouridine as the 3' end group. Enzymatic hydrolysis of this product with snake venom diesterase and bacterial alkaline phosphatase revealed only 1-β-D-arabinosyluracil and uridine; no xyloside was detected. While the primary product containing O^2, 2'-cyclouridine was practically undegraded by these two enzymes, the shorter oligoarabinouridylyl-3'uridylates were completely hydrolyzed. The hydrolysis of the longer chains was a slow process and did not always go to completion. The treatment of fractions larger than trinucleotides with spleen phosphodiesterase always resulted in incomplete hydrolysis. Nevertheless, the presence of linked ara-

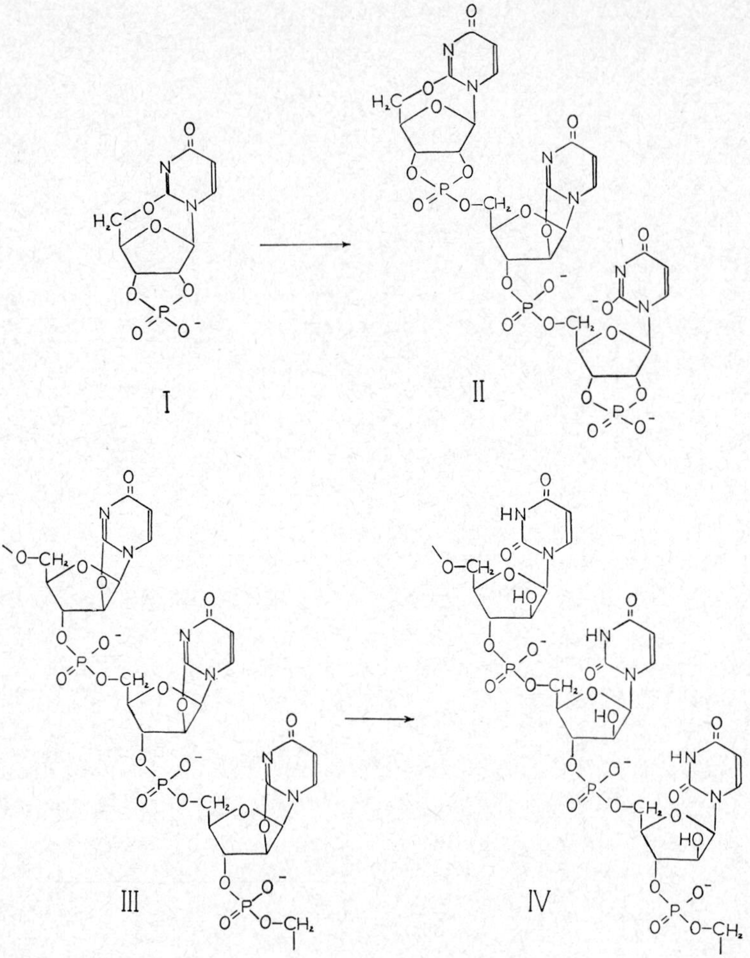

Scheme 1

binouridylates in the reaction product is compatible only with the predominance, if not exclusiveness, if $3' \rightarrow 5'$ phosphodiester linkages.

Several interesting properties of oligoarabinonucleotides have already been
published. The stability of the interarabinonucleotidic bond to both acid
and alkali was first noted by Wechter [10]. The inhibition of several enzymes
by polyarabinouridylates was discovered by Schramm and Ulmer-Schurnbrand [11], and is described in this volume. While we have shown the competitive nature of ribonuclease inhibition, we have not observed any significant binding of Phe-tRNA to E. coli ribosomes in the presence of our larger

polyarabinouridylates [12], in contrast to the findings of these authors. While this discrepancy may be due to the differences in the purity of the two polymers which have been obtained by different methods, it may also be explained by the different Mg^{++} concentrations employed in the binding studies. More significantly, we could not observe any polypeptide synthesis, and, thus the messenger role of polyarabinonucleotides seems to be ruled out. In agreement with this, the mixing curve of polyarabinouridylic acid with polyadenylic acid in 0.2 M NaCl – 0.01 M Mg at 8 °C was found to be a straight line, indicating that no double strand has been formed. We have also found that the optical rotation $[\alpha]_D^{20}$ of oligoarabinouridylyl –(3′) ribouridylates is a linear function of their arabino content. These data suggest that polyarabinouridylic acid may be the structurally most disordered polynucleotide, i.e. devoid of the usual stacking interactions. The molecular model of the dinucleotide

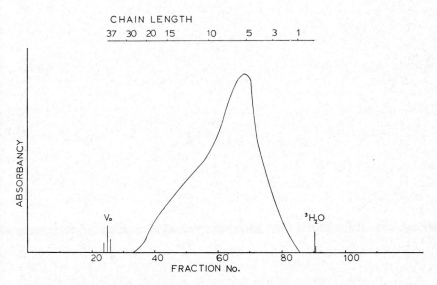

Fig. 2. Distribution of polymerization products from 5′-O-mesyl uridine-2′, 3′-cyclic phosphate on Sephadex G-75 column (90 × 1.6 cm) in 0.01 M triethylammonium bicarbonate, pH 8.5; 3 ml fractions.

illustrates the shielding effect of the C-2′-OH on the uracil bases, which might prohibit the stacking.

One of the most interesting aspects of the chemistry of polyarabinouridylate formation is that, while the final step is irreversible, the primarily formed poly-O^2, 2'-cyclouridylic acid can be in part reverted to the riboderivative through the cyclic triester intermediate. This acid–base catalyzed equilibrium is currently under study.

The displacement of sulfonates. The use of sulfonates as leaving groups is quite common in carbohydrate and nucleoside chemistry, yet there is only one short report on their utilization for oligonucleotide synthesis [13]. The displacement of a 5' sulfonate by a nucleotide appears particularly attractive, because the great difference in the nucleophilicities between the phosphate and water might render the system less sensitive to moisture, and therefore, a better ratio of phosphorylation to activation could be achieved.

With the purpose of obtaining polyarabinouridylic acid by yet another simple method, we have studied the polymerization of 5'-O-mesyl uridine-2', 3'-cyclic phosphate. The reaction, which was carried out in 0.5 M dimethylformamide solution in the presence of tri-n-butylamine at 100 °C for 5 days, showed a surprisingly high degree of polymerization (fig. 2). As in the polymerization of O^2, 5'-cyclouridine-2', 3'-cyclic phosphate, the labile triesters were formed first. The primary product consisted of a copolymer of uridine and O^2, 5'-cyclouridine, the ratio of which reflected the amount of water originally present. Alkaline hydrolysis and subsequent elimination of the terminal sulfonate and phosphate groups gave short oligoarabinonucleotides with ribouridine at the 3' end. Since our best conversion so far has remained under 50%, the yield dropped sharply after the tetranucleotide fraction, in the course of separation on a DEAE column. Thus this method is especially suitable for the preparation of di-, tri- and tetra-arabinouridylates.

Prebiotic significance of the ribo to arabino conversion. Already there are numerous ways to synthesize polyarabinonucleotides of the pyrimidine bases under possible prebiotic conditions. The very significant discovery of Walwick et al. [14] of 1959 clearly demonstrated the convertibility of the pyrimidine ribonucleotides into arabinonucleotides in polyphosphoric acid; i.e. a potential activating agent. In 1967 we reported the partial conversion of polyribouridylic acid into the heteropolymeric poly (ribo, arabino) uridylic acid by thermal activation in the presence of certain activating agents [15]. Two more de novo syntheses were presented in this article; and another distinctly different method was developed in Prof. Schramm's laboratory [11], which is also discussed in this volume. Several other approaches are feasible, some of them run at room temperature, which are now under investigation in our laboratory.

On the basis of this variety of straightforward syntheses and for a number of other reasons, it is tempting to suggest that polyarabinonucleotides had once existed in prebiological times prior to the formation of polyribonucleotides. More precisely, it seems possible that the prebiological evolution of polynucleotides had its starting point in the polyarabinonucleotides of the pyrimidine bases.

One major point of argument is the great difference in the thermal stabilities of polyribonucleotides and polyarabinonucleotides. For example, while the former ones are degraded in hours in solution, even close to neutrality, in the vicinity of 100 °C, polyarabinouridylic acid remains unchanged under these conditions. A further support of our hypothesis can be found in the convertibility of the ribo moiety to the arabino moiety at the monomer and polymer level by the action of a variety of activating agents. Indeed, it seems that most conceivable methods of polymerizing uridylic acid at temperatures above 60 °C lead to the inversion of the configuration at the C-2'. This observation makes the temperature requirement for the prebiogenesis of ribonucleic acid particularly pointed.

Although some properties of polyarabinouridylic acid are surprising, it is fitting from an evolutionary point of view to find that the possibly most primitive polynucleotide possesses the highest degree of disorder structurally and the lowest capacity to function as a template. Since all ramifications of the polyarabinouridylate formation are not yet clear, and even less clear is their prebiological significance, these polymers will undoubtedly form the subject of further research and, of course, speculation.

Significance of phosphorylation via displacement on carbon. From a preparative point of view it appears that the study of displacement reactions, particularly the cyclonucleoside method, will lead to useful contributions in the chemical synthesis of polynucleotides. The great potentiality of this approach for obtaining polymers of high molecular weight is particularly striking when the present lack of advanced expertise is considered.

Although it is a matter of opinion, we feel that reactions of this type which are run at elevated temperatures, have some relevance to the prebiotic formation of polynucleotides in addition to other well recognized means of phosphorylation. Several potential prebiotic leaving groups are feasible, of which the chloride appears the most attractive because of its abundance. Other phosphate related nucleophiles, such as thiophosphates and thiophosphoroamidates, could also be considered; they could have been formed under possible 'primitive earth' conditions and possess a superior nucleophilicity. In our laboratory we have recently obtained poly-5'-thiodeoxythymidylate in a displacement reaction which is characterized by high efficiency and smoothness of condensation.

References

[1] F. Cramer, Angew. Chem. Intern. Ed. 5 (1966) 172.
[2] D. M. Brown, Advances in organic chemistry, methods and results, R. A. Raphael, E. C. Taylor and H. Winberg, eds. (Interscience, New York, 1963); also, H. G. Khorana, Some recent developments in the chemistry of phosphate esters of biological interest (John Wiley and Sons, New York, 1961).
[3] R. L. Letsinger and K. K. Ogilvie, J. Am. Chem. Soc. 89 (1967) 4801; also Letsinger et al., J. Am. Chem. Soc. 89 (1967) 7146.
[4] D. T. Elmore and A. R. Todd, J. Chem. Soc. (1952) 3681.
[5] H. G. Khorana, Can. J. Chem. 32 (1964) 227.
[6] A. M. Michelson, The chemistry of nucleosides and nucleotides (Academic Press, London and New York, 1963) p. 15.
[7] Y. Mizuno and T. Sasaki, J. Am. Chem. Soc. 88 (1966) 863 and footnotes cited therein.
[8] A. M. Michelson, J. Chem. Soc. (1962) 979.
[9] T. Hohn and H. Schaller, Biochim. Biophys. Acta 138 (1967) 466.
[10] W. J. Wechter, J. Med. Chem. 10 (1967) 762.
[11] G. Schramm and I. Ulmer-Schurnbrand, Biochim. Biophys. Acta 145 (1967) 7.
[12] J. Nagyvary, R. Provenzale and J. M. Clark, Biochem. Biophys. Res. Commun. 31 (1968) 508.
[13] J. Nagyvary, Abstracts, 150th National Meeting of the American Chemical Society, Atlantic City, N. J. (Sept. 1965) p. 87C.
[14] E. R. Walwick, W. K. Roberts and C. A. Dekker, Proc. Chem. Soc. (1959) 84.
[15] J. Nagyvary, Abstracts, XXI. Congress of IUPAC, Prague (Sept. 1967) p. N-48.

POLYMERIZATION OF DEOXYRIBONUCLEOTIDES
BY ULTRAVIOLET LIGHT

J. H. McREYNOLDS, N. B. FURLONG, P. J. BIRRELL,
A. P. KIMBALL and J. ORÓ

Department of Biophysical Sciences, University of Houston, Houston, Texas, U.S.A.

*and Department of Biochemistry, University of Texas, M.D. Anderson Hospital
and Tumor Institute, Houston, Texas, U.S.A.*

Plausible mechanisms have been proposed for the abiotic origin of purines, pyrimidines and sugars [1–4]. Other investigations have also shown that nucleosides and nucleotides can be formed from these precursors [5], under primitive earth conditions.

This work is concerned with the last step in the prebiological origin of nucleic acids, namely the condensation of nucleotides into polymers which might serve as templates for their own self-replication or direct the condensation of amino acids to form protoenzymes. Since reducing conditions are assumed to have prevailed before the advent of photosynthetic organisms, this requires that the first nucleic acids be of the deoxy-series due to the hydrolysis of ribonucleic acids containing a $2'$-hydroxyl group in a basic environment.

In studying the results of ultraviolet light on DNA polymerase, Furlong et al. [6] observed that irradiation of nucleotides, singly or in mixtures, consistently resulted in the formation of polymeric material. The conditions employed would be plausible on the primitive earth, that is, a dilute, slightly basic, aqueous solution at moderate temperatures [1]. In addition, the lower atmospheric oxygen levels would allow ultraviolet light of the proper intensity and wave length to reach the surface of the earth [7].

This work describes our experiments on the irradiation of labelled deoxyribomononucleotides by ultraviolet light and our attempts to further characterize the products of this reaction.

Materials and methods

Irradiation. The composition of a typical reaction mixture is given in table 1. The ^{32}P nucleotides, isolated from *E. coli* grown in a medium containing

TABLE 1

Typical reaction mixture

Phosphate buffer, 0.05 M, pH 7.3	15 μl
MgCl$_2$, 0.2 M	10 μl
dAMP, 0.001 M to 0.1 M	25 μl
dAMP-8-^{14}C, 10 μC/ml	10 μl
or	
dAT^{32}P, dGT^{32}P, dCT^{32}P or TT^{32}P	50 μl
H$_2$O	to 250 μl Total

$K_2H^{32}PO_4$, were used in one of the first experiments. In most of our work we used dAMP-8-^{14}C with an activity of 10 μC/ml. Solutions were irradiated with nitrogen stirring in quartz cuvettes with a 2 mm path length for periods up to two hours. A "Pen-Ray" quartz lamp manufactured by Ultraviolet Products, Inc. was used at a distance of 3 cm from the cuvette. A total dose of approximately 10^6 erg/mm^2 was delivered in 90 min at 253.7 mμ [8]. This was determined using a Westinghouse wattmeter at Baylor University College of Medicine, Houston, Texas [8].

Assay procedure. At selected time intervals during the irradiation, 25–50 μl aliquots were withdrawn and spotted at the origin of a sheet of DEAE cellulose paper (Whatman DE-81). The chromatograms were developed by descending chromatography in 0.75 M ammonium bicarbonate for the separation of oligonucleotides according to chain length [9].

After drying, the origins were cut out and counted or the entire chromatogram may be autoradiographed. Alternatively duplicate aliquots were spotted on a 1 cm filter paper disc which had been treated with 25 μl of a 0.1 M solution of sodium pyrophosphate at least 30 min before the experiment. One disc was allowed to dry undisturbed as a measure of the total radioactivity per aliquot. The other discs were given three successive washings in cold 5% TCA and a final wash in 95% ethanol. After drying the discs were counted for the acid insoluble radioactivity in a liquid scintillation counter with a toluene scintillation solution. This is a modification of the method described by Bollum for the assay of DNA polymerase activity [10].

Enzymatic assays. Solutions were irradiated as described above for periods up to two hours. At the end of this time 50 μl of irradiated solution was spotted on a sheet of DEAE paper. Additional 50 μl aliquots were placed in test tubes along with buffer and enzyme. After incubation, 50 μl aliquots were with-drawn from each tube and spotted on the same sheet of DEAE paper. The origins of the developed chromatograms were cut out and counted in toluene scintillation solution. The conditions of incubation and the contents of each tube are given below:

Control. 50 μl irradiated solution; 10 μl, 5 mg/ml heat denatured DNA; 10 μl 1 M tris-HCl, pH 8.8, 10 μl H_2O; incubated at 37 °C for one hour.

DNAse I. Same as control except 10 μl, 1 mg/ml DNAse I substituted for H_2O; incubated at 37 °C for 20 min.

Venom phosphodiesterase. As above except 10 μl, 1 mg/ml venom phosphodiesterase (Russel's Viper Venom, Calbiochem, B grade); incubated at 37 °C for two hours.

Spleen phosphodiesterase A. 50 μl irradiated solution; 10 μl DNA; 10 μl, 1 mg/ml spleen phosphodiesterase (Bovine pancreas, Calbiochem, B grade); incubated at 37 °C for one hour.

Spleen phosphodiesterase B. Same as A plus 1 μl, 1 mg/ml alkaline phosphatase (Miles Laboratory); incubated at 37 °C for one hour.

Chain length determination. The chain length of polymers incorporating ^{32}P nucleotides was determined by the method of Furlong [9]. Total radioactivity is counted and the terminal phosphate cleaved by alkaline phosphate is extracted with molybdic acid and heptanol and counted. Chain length was determined as the ratio of total to terminal phosphate.

Results

Fig. 1 and 2 show autoradiographs of two experiments with dATP-α-^{32}P and dAMP-8-^{14}C. In both cases the material at the origin increases with time, while the monomer spot decreases in intensity. These results indicate that the material at the origin may be formed from the intact nucleotide since both the purine moiety and the phosphate group are incorporated.

Fig. 3, showing the results of 25 experiments, demonstrates that the amount of polymer produced is proportional to the dose of irradiation.

The kinetics of the reaction is shown in fig. 4. For this experiment aliquots were taken from a 500 μl reaction mixture every five minutes.

The reaction was carried out at different temperatures. These results appear in fig. 5. It is apparent that there is no temperature dependence indicating that the rate of polymerization is limited by the concentration of a species produced photochemically.

The reaction is concentration dependent with the optimum yield occurring at a nucleotide concentration of 5×10^{-4} M. Fig. 6 shows the results obtained using this concentration which gave yields of 5%, 12% and 17% incorporation at 30, 60 and 90 min, respectively. The corresponding values

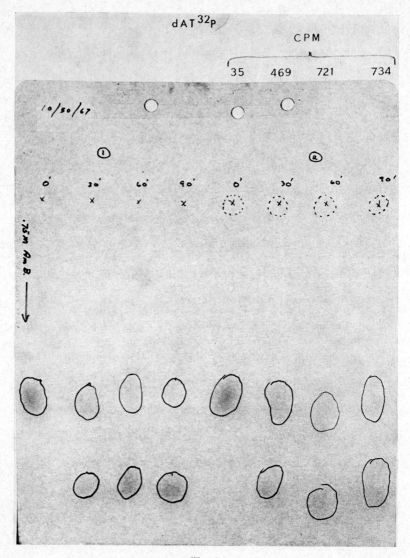

Fig. 1

for more or less concentrated solutions were lower, e.g. the yields for a 10^{-4} M solution were only 0.3%, 1.2% and 1.5%, respectively.

Enzymatic degradation was used to verify the nucleic acid nature of the polymer and elucidate the types of linkages formed. Incubation with DNAse I showed degradation of 28.6% and 71.3% with dAMP-8-^{14}C and dGTP-U-

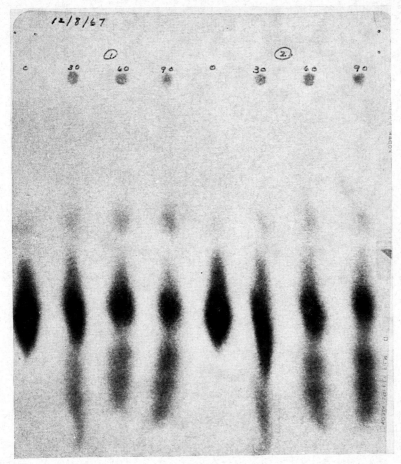

Fig. 2

^{32}P used as substrates. The results of this experiment are shown in table 2. A second degradation with DNAse I of a dAMP product gave 46.2% degradation. Venom phosphodiesterase, which attacks the 3' linkages, gave degradations of 22.6% and 50.4% in two experiments with dAMP. Since the substrate has a 5' phosphate, these results may be interpreted as the percentages of normal 3'-5' linkages. Incubation with spleen phosphodiesterase degraded only 16.8% of the polymer. Since this enzyme requires a free 5' hydroxyl group, this is an indication of the amount of 5' phosphate removed by the ultraviolet light. The addition of alkaline phosphatase to spleen phosphodiesterase provided a free 5' hydroxyl group. Incubation of the product with

TABLE 2

DNAse digestion

Experiment	Cpm at origin	
	dAMP-8-^{14}C	dGT^{32}P
1. Control (no UV)	5.6	2180.0
2 hr UV	33.5	5789.0
2. Control + DNAse I	6.2	1312.0
2 hr UV + DNAse I	23.8	1661.0

Results of two separate experiments with DNAse I aliquots of control and irradiated solution were chromatographed before and after incubation with DNAse I as described in the text.

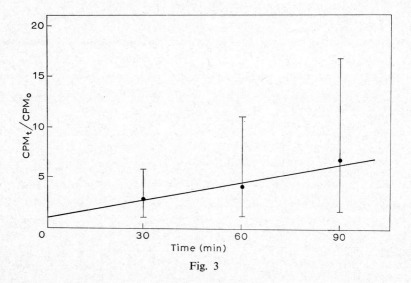

Fig. 3

both of the enzymes gave a degradation of 19.6%. In all the enzyme experiments the degradation is determined by the difference in counts at the origin between the control and enzyme incubations. The degradation in control incubations was small in all cases.

The chain length determinations carried out showed chain lengths of 12.9 and 10.2 units with dGTP and TTP respectively, as illustrated in table 3.

Discussion

The results presented indicate that ultraviolet light is responsible for the in-

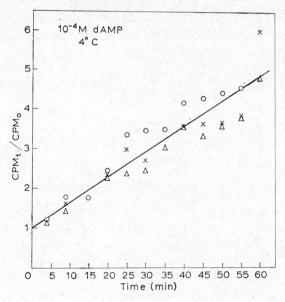

Fig. 4

Effects of temp.

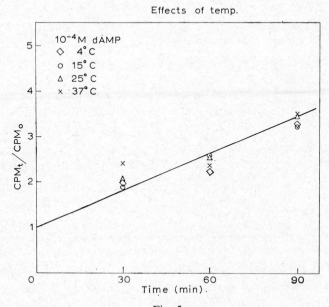

Fig. 5

TABLE 3

Chain length determinations

Product	Chain length (units)
dGT^{32}P + 2 hr UV	12.9
Control no UV	2.96
TT^{32}P + 2 hr UV	10.21
Control no UV	1.31

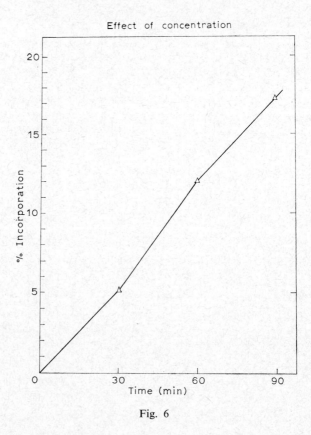

Fig. 6

corporation of intact dAMP into an acid insoluble product. That the substrate is incorporated intact is strongly suggested by the similarity of results
with the labels in the adenine or the phosphate. In other experiments, with
more energetic UV, we have found that decomposition of adenine produces
4-amino-5-imidazolecarboxamide (AICA) as the major identifiable species.

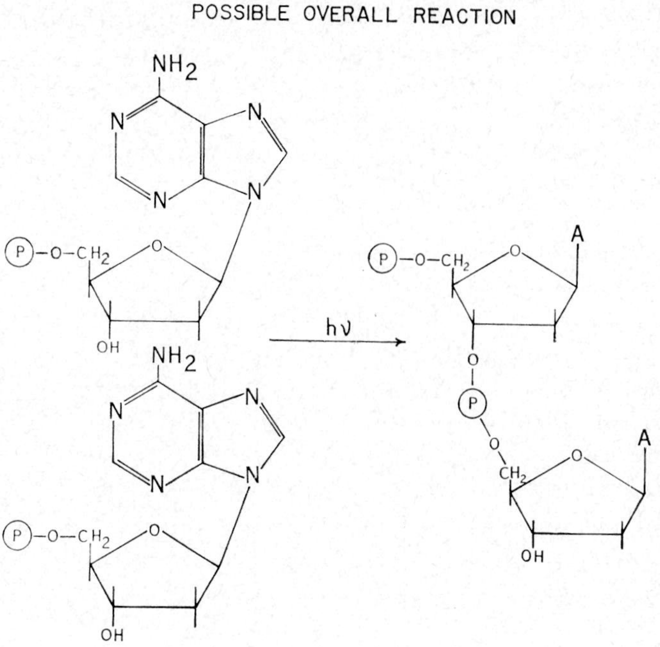

POSSIBLE OVERALL REACTION

Fig. 7

This compound appears as a yellow spot near the solvent front in our chro-matography system. AICA was not detected in any experiments done with the smaller Pen-Ray lamp which further supports our belief that dAMP is in-corporated intact. The linear increase of the product with time, together with the absence of any temperature dependence point toward a photo-chemically initiated mechanism.

While we are not in a position to examine the mechanism of the reaction at this time, we can argue the plausibility of the results that have been found. The sugar and phosphate groups which we would like to implicate in the polymerization do not absorb at the wavelength used. The heterocyclic bases however have their maximum absorption near 257 mμ. It has been found in other systems [11] that excited triplet states may be stabilized by the excited singlet state of oxygen. This may extend the lifetime of the triplet state sufficiently to allow its participation in inter- or intramolecular energy transfer processes resulting in the localization of energy in the bond where polymerization occurs. This type of mechanism applied to our system could account for the transfer of energy from the purine to the sugar and phosphate groups of dAMP.

The degradation of the product by DNAse I and venom phosphodiesterase is consistent with, but by no means proves, the hypothesis that our product is similar to naturally occurring DNA. First, we have yet to show unequivocally that the enzymatic decomposition products are the naturally occurring monomers, although paper chromatography (this work and that of Furlong et al. [6]) shows that these products are undistinguishable from the natural monomers. Secondly, the action of the enzymes used in these experiments on polymers with abnormal linkages has not been determined. We are not aware of any published data on this problem, however, we are planning a study of these enzymes as a separate problem.

Another question to be answered is the absence of the expected chain length distribution. Our radioautographs show the material at the origin, which should consist of polymers with chain lengths over 20, apparently exceeding the concentration of dimers, trimers and other oligomers not retained at the origin. One would expect a decreasing yield with increasing chain length tending toward a maximum chain length. An explanation of this apparent inconsistency is not attempted at this time as it would necessarily be more speculative than factual. Instead we will only report that we are attempting to establish with more confidence the actual chain length distribution by column chromatography before answering the above question.

In summary our present evidence is such that we tentatively conclude that we are dealing with a synthetic oligo- or polynucleotide. Final proof depends on the isolation of the product in sufficient quantity to apply more extensive analytical tests.

Acknowledgements. This work was made possible by the following grants: NGR-005-002 and NsGT-52-Sup.4 from the National Aeronautics and Space Administration, GM-00940-06 from the Public Health Service, 68-04551 from the National Defense Education Act, G 120 from the Robt. A. Welch Foundation and IN-43-H7 from the American Cancer Society. In addition we greatfully acknowledge the technical assistance of Elmer Scheltgen and Patrick Willis.

References

[1] J. Oró, Stages and mechanisms of prebiological organic synthesis, in The origin of prebiological systems and of their molecular matrices, S. W. Fox, ed. (Academic Press, New York, 1965) p. 137.
[2] R. A. Sanchez, J. P. Ferris and L. E. Orgel, Science *154* (1966) 184.

[3] N. W. Gabel and C. Ponnamperuma, Nature 216 (1967) 453.
[4] C. Reid and L. E. Orgel, Nature 216 (1967) 455.
[5] A. Beck, R. Lohrmann and L. E. Orgel, Science 157 (1967) 952.
[6] N. B. Furlong et al., Proc. Am. Cancer Res. 8 (1967) 20.
[7] L. V. Berkner and L. C. Marshall, Proc. Natl. Acad. Sci. U.S. 53 (1965) 1215.
[8] E. Scheltgen and J. Oró, unpublished observations.
[9] N. B. Furlong, Methods in enzymology, nucleic acids, Vol. 12, L. Grossman and K. Moldave, ed. (1967) 318.
[10] F. J. Bollum, J. Biol. Chem. 234 (1959) 2733.
[11] S. P. McGlynn, F. J. Smith and G. Cilento, Photochem. Photobiol. 3 (1964) 269.

RECENT ADVANCES IN STUDIES OF EVOLUTIONARY RELATIONSHIPS BETWEEN PROTEINS AND NUCLEIC ACIDS*

THOMAS H. JUKES

University of California,
Space Sciences Laboratory, Berkeley, California, U.S.A.

The diversity of living organisms has been a subject of interest to human beings for many years. At one time it was customary to ascribe the origin of the so-called lower forms of life to spontaneous generation. This theory was destroyed by the experiments of Spallanzani and Pasteur, but the theme of a special creation continued in many circles to be reserved for man himself. About one hundred years ago, the publication of the *Origin of Species* led to wide acceptance of the idea that all life had a common origin; this was mentioned even in "The Mikado" by Gilbert and Sullivan. The concept of divergent evolution has been reinforced by an increasing body of evidence, including taxonomy and systematics, ecology, paleontology, biochemistry, Mendelian genetics, and mutations. It leads back to a single organism. We shall explore this concept in terms of deoxyribonucleic acid (DNA).

The formula for DNA proposed by Watson and Crick [1] showed that the molecular basis of heredity rests on the sequences of four bases; adenine, cytosine, guanine and thymine (A, C, G and T) in DNA molecules. This provides a direct key to the molecular mechanism of evolution. Since all hereditary information is carried as linear sequences of four variables, evolution can take place only through changes in such sequences, consisting of repetitive changes, molecular shortening, and replacement of one base with another in DNA. These alterations produce phenotypic changes in the organism. Most of these changes occur in proteins. The net result is change in the fitness of the organism for its environment. This leads to natural selection: the emergence of some species, and the extinction of others.

The new field of molecular evolution measures the evolutionary changes in DNA molecules by comparing their base sequences. One method for comparing them is by the annealing procedure, in which the ability of single strands of DNA from two different species to form hybrids is quantitatively measured. Another is the indirect method of determining and comparing the amino acid sequences in proteins. In a very few cases, it has been possible

* Supported by NASA grant NsG 479 to the University of California, Berkeley. Reprinted by permission from Space Life Sciences *1* (1969) 469.

to analyze and directly compare the base sequences in molecules of RNA.

It is fortunately possible, although the procedures are laborious, to determine the amino acid sequences of proteins. The genetic code enables the sequences to be translated back into the base sequences in DNA that code for the proteins. There are some limitations to this procedure because of certain ambiguities in the code. However, the method is sufficiently accurate to provide a lot of information on the evolution of proteins. Such information is deduced from comparisons of proteins that have identical or analogous functions, and similar but not identical sequences. Such proteins are obtained in most cases from different species of organisms.

To re-emphasize the preceding, the information for heredity and hence for evolution is carried as a linear sequence of four variables; the bases in the two strands of DNA molecules.

Changes in the length and ordering of the sequence are responsible for all inherited changes, and hence for the *genotypic* component of evolution.

The other two controlling forces in evolution are *phenotypic expression*, which is the direct result of translation of the information in the sequence, and *natural selection*, which is exerted on the phenotype.

So far, the DNA model has been completely adequate for the study of genetics and evolution. There is no need to propose any other model until the DNA system has been found wanting. The chances of this are vanishingly small.

DNA is essentially inert. It functions solely as a repository for information. Its physical and physico-chemical properties are of the greatest interest and importance, because they have a bearing on the manner in which it is replicated and transcribed, but they are not properties that we associate with living organisms. The *transcription* of DNA is defined as the production of complementary RNA copies of one of the two DNA strands. All transcription is carried out by the action of an enzyme, RNA polymerase. The length of the single-stranded RNA molecules that are produced by transcription is controlled by *start* and *stop* signals on the DNA molecules. Presumably these signals, like all other information in DNA, are linear permutations with specific sequences. Recent evidence is that the *start* signal consists of a short sequence of C's or, at least, a C-rich cluster [2].

There is no evidence that modified bases in DNA, such as methyl cytosine in plant DNA, are needed for coding purposes.

RNA and its functions

The coding properties of DNA are studied by analyzing the structure and function of RNA. Three main functional classes of RNA are recognized: messenger, transfer and ribosomal. Messenger RNA consists of long single strands of RNA, containing unmodified bases. Their function is to specify amino acid sequences in polypeptides. The code in messenger RNA molecules (table 1) consists entirely, or almost entirely, of 64 codons, each consisting of one of the 64 three-letter permutations of A, C, G and U (uracil).

TABLE 1

The genetic code

UUU Phenylalanine	CUU Leucine	AUU Isoleucine	GUU Valine
UUC Phenylalanine	CUC Leucine	AUC Isoleucine	GUC Valine
UUA Leucine	CUA Leucine	AUA Isoleucine	GUA Valine
UUG Leucine	CUG Leucine	AUG Methionine	GUG Valine
UCU Serine	CCU Proline	ACU Threonine	GCU Alanine
UCC Serine	CCC Proline	ACC Threonine	GCC Alanine
UCA Serine	CCA Proline	ACA Threonine	GCA Alanine
UCG Serine	CCG Proline	ACG Threonine	GCG Alanine
UAU Tyrosine	CAU Histidine	AAU Asparagine	GAU Aspartic acid
UAC Tyrosine	CAC Histidine	AAC Asparagine	GAC Aspartic acid
UAA Chain Termn.	CAA Glutamine	AAA Lysine	GAA Glutamic acid
UAG Chain Termn.	CAG Glutamine	AAG Lysine	GAG Glutamic acid
UGU Cysteine	CGU Arginine	AGU Serine	GGU Glycine
UGC Cysteine	CGC Arginine	AGC Serine	GGC Glycine
UGA Chain Termn.	CGA Arginine	AGA Arginine	GCA Glycine
UGG Tryptophan	CGG Arginine	AGG Arginine	GGG Glycine

The transfer RNA (tRNA) molecules are more complex than those of messenger RNA. They double back on themselves to form helical regions and loops. One of the loops contains an anticodon of 3 consecutive bases, which pairs with a messenger RNA codon. It is possible that there are not more than 55 anticodons because of ambiguous pairing in the third base position of the codons. Seven anticodons have been identified by analyzing tRNA molecules.

There are about 75 to 85 nucleotides in a tRNA molecule. The molecule is first formed as a sequence of A, C, G and U, following which, some of these bases are changed by enzymes into modified bases, such as dimethyl guanine, etc. TRNA molecules have a distinctive and complex three-dimensional structure. Each tRNA molecule contains at least four regions with specific functions: first, the anticodon: next, the sequence TΨCG or TΨCA; which may bind tRNA to a ribosome; third, an unidentified sequence that is specific for the enzyme which attaches one, and only one, of the 20 amino acids to the end of the molecule; and, fourth, the terminal -CCA for the attachment of the amino acid, which is the first step in translation of the code.

We must conclude that the short length of the tRNA molecule and its base composition enable the cell and its enzyme systems to distinguish it from other RNA molecules, such as those of messenger RNA.

The third class of RNA molecules are the ribosomal RNAs. These molecules probably do not contain information for protein synthesis. The two larger ribosomal RNA molecules contain some modified bases, but the smaller, third one, does not. The function of ribosomal RNA is to combine with several specific proteins to form a ribosome, which is about 20 to 30 mμ in diameter. Proteins are synthesized on the surface of ribosomes, and ribosomes participate in the pairing reaction between codons in messenger RNA and anticodons in transfer RNA. This pairing leads to the selection of a specific sequence of amino acids by a messenger RNA molecule. This sequence is coupled together by peptide linkages to form a protein such as insulin, hemoglobin or casein. There are thousands of different proteins in each of the higher species of living organisms.

Hydrogen bonding

Hydrogen bonding in the nucleic acids was discovered by Watson and Crick [1]. They proposed that two hydrogen bonds between each AT and GC pair held the two strands of DNA molecules together. Later, Pauling and Corey [3] pointed out that the GC pair had 3 hydrogen bonds. Hydrogen-bonded strands of DNA and RNA in helices always run in opposite directions as defined by the sugar-phosphate linkages in the nucleid acids.

Hydrogen-bonding in RNA is usually by means of AU and GC pairs. As a result of the nature of the RNA polymerase reaction, RNA occurs in single strands and these can double back on each other to form two-stranded helical regions, as in transfer RNA.

Hydrogen bonding takes place between the codon of messenger RNA and the anticodon of transfer RNA in the selection of amino acids in protein synthesis. The results of this hydrogen bonding reaction are included in a group of relationships which is termed the amino acid code.

It was proposed by Crick [4] that codon–anticodon pairing includes a special type of hydrogen bonding, as follows: the first two bases of the codon and the second two bases of the anticodon pair according to the Watson–

Crick rule, but there is a certain amount of play or "wobble" between the third base of the codon and the first base of the anticodon. The pairing in this scheme is shown in table 2.

This complex pattern is at the root of the translation of the genetic code. Apparently, it is Nature's device for apportioning the codes for 20 amino acids among the 64 possible permutations of the 4 bases in messenger RNA

TABLE 2

Codon – anticodon pairing: the wobble hypothesis*

Anticodon	Anticodon
$-$ I $-$ G $-$ C $- \rightarrow$	$-$ G $-$ U $-$ A $- \rightarrow$
C ‖‖ ‖‖	⎰ ‖ ‖
$- \wr -$ C $-$ G $- \leftarrow$	$-$ C $-$ A $-$ U $- \leftarrow$
$-$ U $-$	$-$ U $-$
$-$ A $-$	
Codon	Codon

Anticodon	Anticodon
$-$ C $-$ A $-$ A $- \rightarrow$	$-$ U $-$ U $-$ U $- \rightarrow$
‖‖ ‖ ‖	⎰ ‖ ‖
$-$ G $-$ U $-$ U $- \leftarrow$	$-$ A $-$ A $-$ A $- \leftarrow$
	$-$ G $-$
Codon	Codon

* I = inosine.

and for providing two anticodons to translate each codon. It is not known whether transfer RNAs carrying all of the possible anticodons are present in all organisms: in fact, it may well be that some are missing from some species. The "wobble" hypothesis implies that anticodons with A in the first position do not occur, and so are not used in the translation of the genetic message. Presumably, the A occurring in this position is deaminated to hypoxanthine by a specific enzyme during the modification of transfer RNAs subsequently to their transcription. Hypoxanthine riboside is termed inosine, and hypoxanthine pairs preferentially with C. However, in the "wobble" position it pairs with U or A as well as C. Table 3 shows the known, indicated and predicted anticodons.

Protein homology and evolution

In studying evolution, the genetic code is used to compare the amino acid sequences in two homologous proteins. An example would be a comparison

TABLE 3

The known (double underlining), indicated (single underlining) and predicted anticodons

	IAG Leu	IAU Ile	IAC Val
GmAA Phe	GAG "	GAU "	GAC "
UAA Leu	UAG "		acUAC "
CAA "	CAG "	CAU Met	CAC "
IGA Ser	IGG Pro	IGU Thr	IGC Ala
GGA Ser	GGG "	GGU "	GGC "
UGA "	UGG "	UGU "	UGC "
CGA "	CGG "	CGU "	CGC "
GUA Tyr	GUG His	GUU Asn	GUC Asp
UUA End	UUG Gln	UUU Lys	tUUC Glu
CUA "	CUG "	CUU "	CUC "
	ICG Arg		ICC Gly
GCA Cys	GCG "	GCU Ser	GCC "
	UCG "	UCU Arg	UCC "
CCA Trp	CCG "	CCU "	CCC "

The known anticodons have been identified in tRNA molecules; the indicated anticodons have been identified by binding of tRNA on ribosomes with polynucleotides of known sequence and the remainder of the anticodons are predicted by the wobble hypothesis.

between the hemoglobins of a fish and a mammal. A hemoglobin molecule contains two different polypeptide chains, each with about 140 to 150 amino acids. The α chains of human and carp hemoglobin are identical in about half of the amino acids. It is also possible to compare the hemoglobins in a single species, for there are several different hemoglobins in each vertebrate species, such as human beings, and the chains may be compared with each other to study the manner in which the globins (hemoglobins and myoglobins) have diverged from a common origin, which is termed an archetypal gene.

An example of this procedure is shown in table 4, in which segments of four different globin peptide chains, each corresponding to about 11% of a hemoglobin chain, are compared with each other. The first segment is taken from "muscle hemoglobin", or myoglobin, the next from the α chain of hemoglobin; the third from the β chain and the fourth from the γ chain. All these chains are present in every human being; the γ chain being present prior to birth.

TABLE 4

Amino acid sequences in portions of globin chains. The numbering is from the beginning of all four chains, including gaps

<u>79</u>	<u>94</u>

M	-Lys-Lys-Gly-His-His-Glu-Ile-Glu-Leu-Lys-Pro-Leu-Ala-Gln-Ser-His-
α	-His-Val-Asp-Asp-Met-Pro-Asn-Ala-Leu-Ser-Ala-Leu-Ser-Asp-Leu-His-
β	-His-Leu-Asp-Asn-Leu-Lys-Gly-Thr-Phe-Ala-Thr-Leu-Ser-Glu-Leu-His-
γ	-His-Leu-Asp-Asp-Leu-Lys-Gly-Thr-Phe-Ala-Gln-Leu-Ser-Gly-Leu-His-

M = myoglobin;   α = α hemoglobin;   β = β hemoglobin;   γ = γ hemoglobin.

All four of the globin chains are structurally and chemically related; they are of almost equal lengths, and they coil themselves into convoluted globular molecules of almost exactly the same shape, as shown by crystallography. Each of them holds in its folds a unit of heme, the oxygen-carrying pigment. A striking difference is found, however, in the sequences of amino acids. Myoglobin differs from each of the other three at about 75% of the amino acid sites. The α chain differs from the β and γ chains by about 55%. Finally, the β and γ chains resemble each other more closely than they do either of the other two chains, and differ by only 27%. This state of affairs is just what would be expected if all four of the polypeptide chains had a common ancestor from which they have diverged by a series of widely-spaced events of separation followed by differentiation. The mechanism for this is straightforward in terms of DNA and the genetic code. In table 5 the same sequences of amino acids are shown in terms of possible codons representing them, just as they would be found in the regions of the DNA molecule that are responsible for carrying the genetic information for the globins. The minimum base differences are also shown. Table 6 shows that the β and γ chains separated from each other subsequently to the time when human beings and fishes had a common ancestor, but before the lines of descent separated that led to

TABLE 5

The sequence in the globin chains in table 4 written in the form of their genes

<u>79</u>   <u>94</u>

M A-A-R-A-A-R-G-G-N-C-A-Y-C-A-Y-G-A-R-A-T-Y-G-A-R-T-T-R-A-A-R-C-C-N-C-T-N-G-C-N-C-A-R-T-C-N-C-A-Y

α C-A-Y-G-T-N-G-A-Y-G-A-Y-A-T-G-C-C-N-A-A-Y-G-C-N-T-T-R-T-C-N-G-C-N-C-T-N-T-C-N-G-A-Y-T-T-R-C-A-Y

β C-A-Y-T-T-R-G-A-Y-A-A-Y-T-T-R-A-A-R-G-G-N-A-C-N-T-T-Y-G-C-N-A-C-N-C-T-N-T-C-N-G-A-R-T-T-R-C-A-Y

γ C-A-Y-T-T-R-G-A-Y-G-A-Y-T-T-R-A-A-R-G-G-N-A-C-N-T-T-Y-G-C-N-C-A-R-C-T-N-T-C-N-G-G-N-T-T-R-C-A-Y

Minimum Differences: M : α 20   α : β 12   β : γ 4   N = A, C, G or T;

                      M : β 20   α : γ 13   R = A or G;

                      M : γ 22   Y = C or T.

TABLE 6

Minimum differences per codon in the relationship between certain globin chains

Comparison	Sites compared	Minimum differences per codon				Average
		None	1	2	3	
Myoglobin: α human	140	35	57	47	1	1.10
Myoglobin: β human	144	36	53	54	1	1.14
Myoglobin: γ human	144	36	57	50	1	1.11
Lamprey: α human	130	45	55	30	0	0.89
α human: γ human	139	59	55	25	0	0.76
α human: β human	139	64	53	22	0	0.70
α human: α carp	140	73	42	25	0	0.66
β human: γ human	146	106	29	10	0	0.34
β human: β horse	146	120	18	8	0	0.23
α human: α horse	142	126	12	5	0	0.16
β human: δ human	146	136	9	1	0	0.08
β human: β M. mulatta	146	139	6	1	0	0.05

human beings and horses, in which the β chains differ by 0.23 minimum base differences per codon.

All the homologous proteins that have been so far studied show relationships that are quite analogous to the one set forth for the hemoglobins. These include, among other, the cytochromes c, the bacterial and plant ferredoxins, the insulins, the fibrinopeptides, and pancreatic ribonuclease. The complete amino acid sequences of spinach and *Scenedesmus* ferredoxins [5, 6] are in table 7. These differ by 0.41 minimum base differences per codon. The

TABLE 7

Sequences of two type 1 (chloroplast type) ferredoxins

```
Sp  NH2Ala-Ala-Tyr-Lys-Val-Thr-Leu-Val-Thr-Pro-Thr-Gly-Asn-Val-Glu-Phe-Gln-Cys-Pro-Asp-Asp-Val-Tyr-Ile-Leu-Asp-Ala-Ala-
            1                       2           1       1   2   2   1   1                   2

Sc  NH2Ala-Thr-Tyr-Lys-Val-Thr-Leu-Lys-Thr-Pro-Ser-Gly-Asp-Gln-Thr-Ile-Glu-Cys-Pro-Asp-Asp-Thr-Tyr-Ile-Leu-Asp-Ala-Ala-

             *Lys    *Met
Sp  -Glu-Glu-Glu-Gly-Ile-Asp-Leu-Pro-Tyr-Ser-Cys-Arg-Ala-Gly-Ser-Cys-Ser-Ser-Cys-Ala-Gly-Lys-Leu-Lys-Thr-Gly-Ser-Leu-
              1       1                                       1                               1   1   1       1   1

Sc  -Glu-Glu-Ala-Gly-Leu-Asp-Leu-Pro-Tyr-Ser-Cys-Arg-Ala-Gly-Ala-Cys-Ser-Ser-Cys-Ala-Gly-Lys-Val-Glu-Ala-Gly-Thr-Val-

Sp  -Asn-Gln-Asp-Asp-Gln-Ser-Phe-Leu-Asp-Asp-Asp-Gln-Ile-Asp-Glu-Gly-Trp-Val-Leu-Thr-Cys-Ala-Ala-Tyr-Pro-Val-Ser-Asp-
      1       2                               2       1       1       2                   1               2

Sc  -Asp-Gln-Ser(Asp,Gln)Ser-Phe-Leu-Asp-Asp-Ser-Gln-Met-Asp-Gly-Gly-Phe-Val-Leu-Thr-Cys-Val-Ala-Tyr-Pro-Thr-Ser-Asp-

Sp  -Val-Thr-Ile-Glu-Thr-His-Lys-Glu-Glu-Glu-Leu-Thr-AlaCOOH
      2           1                       1       2

Sc  -Cys-Thr-Ile-Ala-Thr-His-Lys-Glu-Glu-Asp-Leu-Phe-COOH
```

Sp = Spinach [5]
Sc = Scenedesmus [6]
* = Spinach variant
Identical sites 67; Single-base differences 19; Two-base differences 10; Minimum base differences per codon (MBDC) 0.41.

differences have accumulated during a period of about 350 million years of divergent evolution in the plant kingdom. A comparison of three bacterial ferredoxins is in table 8.

Some sets of homologous proteins have diverged from each other more rapidly than other sets during evolution. A possible explanation for this is that some proteins are subject to greater constraints than others in the relation of their structures to their functions. Such constraints will lead to rigorous natural selection and the rejection of numerous deleterious mutants by lethality. The net result would be that the protein would appear to evolve more slowly. An alternative explanation for the same result might be that some regions of the chromosomes are better protected than others against mutations.

It is often necessary to postulate the presence of "gaps" in aligning homologous sequences. The criteria for this are explained by Cantor [10].

Hybridization of DNA as a measure of evolution

Marmur and Lane [11] showed that after separating two strands of DNA, obtained from *Diplococcus pneumoniae*, by heating, it was possible to reunite them by slowly cooling the mixture. This necessitated the pieces of the strands aligning in exactly the same manner as that in which they were originally held together; for the randomization of the sequence of bases in a bacterial DNA molecule is such that, with few exceptions, each sequence is unique when measured over distance of about 15 or more base pairs. Each region must therefore "find" its original partner before it can reunite. It was also found that a single strand of DNA from one organism could form a double strand with fragments of single-stranded DNA of a closely related organism [12]. The hybrid double strand did not bind together as firmly as did a pair of single strands from the same organism. Many refinements of this procedure were subsequently introduced by Bolton, McCarthy, Hoyer, Britten and Kohne [13–19]. An extensive literature has accumulated in which the quantitative nature of binding between single strands of DNA from different organisms has been used as a measure of taxonomy and evolution. The results are quite concordant with those obtained by other procedures, including comparisons of amino acid sequences in homologous proteins, and also comparisons made by classical methods of taxonomy and systematics. Two new findings have emerged: one is that the DNA of higher organisms contains regions of highly repetitive sequences, some of which are found to have been duplicated 100,000 or more times, and the other is that some portions of

TABLE 8

Bacterial ferredoxins: primary structures and base differences per codon in internal duplication

(i) *Clostridium pasteurianum* (a)

```
1        *         *      10    *    *       *        *      20 *         *    29
Ala-Tyr-Lys-Ile----Ala-Asp-Ser-Cys-Val-Ser-Cys-Gly-Ala-Cys-Ala-Ser-Glu-Cys-Pro-Val-Asn-Ala-Ile-Ser-Gln-Gly-Asp-Ser-
30                                    40                                    50                                    58
-Ile-Phe-Val-Ile-Asp-Ala-Asp-Thr-Cys-Ile-Asp-Cys-Gly-Asn-Cys-Ala-Asn-Val-Cys-Pro-Val-Gly-Ala-Pro-Val-Gln-Glu-- - -
 2   1   2   0    0   0   1   0   1   2   0   0   2   0   0   1   1   0   0   0   2   0   2   2   0   1         Ave.
                                                                                                               0.77
```

(ii) *Clostridium butyricum* (b)

```
Ala-Phe-Val-Ile----Asn-Asp-Ser-Cys-Val-Ser-Cys-Gly-Ala-Cys-Ala-Gly-Glu-Cys-Pro-Val-Ser-Ala-Ile-Thr-Gln-Gly-Asp-Thr
-Gln-Phe-Val-Ile-Asp-Ala-Asp-Thr-Cys-Ile-Asp-Cys-Gly-Asn-Cys-Ala-Asn-Val-Cys-Pro-Val-Gly-Ala-Pro-Asn-Gln-Glu-- - -
 2   0   0   0    2   0   1   0   1   2   0   0   2   0   0   0   2   1   0   0   0   1   0   2   1   0         0.69
```

(iii) *Micrococcus aerogenes* (c)

```
                                              , Lys, Pro,
                                                    Pro-
Ala-Tyr-Val-Ile-- - -Asn-Asp-Ser-Cys-Ile-Ala-Cys-Gly-Ala-Cys(Gln, Gly, Asn, Cys, Pro, Val, Asn  Ala, Ile, Gln, Gln, Gly) - Ser
-Ile-Tyr-Ala-Ile-Ala-Ile-Asp-Ala-Asp-Ser-Cys-Ile-Asp-Cys-Gly-Ser-Cys-Ala-Ser-Val-Cys-Pro-Val-Gly-Ala-Pro-Asn-Pro-Glu-Asp-- -
 2   0   1   0    2   0   0   0   0   1   0   0   1   0   0   1   0   0   2   1   2   0   0   2   0   2   2   1   1         0.77
```

Prototype

```
1                            10                           20                           29
Ala-Phe-Val-Ile-Asp-Ala-Asp-Ser-Cys-Ile-Asp-Cys-Gly-Ala-Cys-Ala-Ser-Val-Cys-Pro-Val-Gly-Ala-Ile-Asn-Gln-Gly-Asp-Ser-
Tyr[†]                                                                                                            Pros   Glus
```

Comparison	Base differences			
	1 base	2 base	total	Per codon
Cl. past./Cl. but.	6	3	12	0.22
Cl. past./Mic. aer.	9	6	21	0.39
Cl. but./Mic. aer.	11	4	19	0.35

* Invariant sites in both halves of molecule. The alignment is that proposed by Tsunoda et al. [7].
† Alternate possibilities in prototype.
References: (i) Tanaka et al. [8]; (ii) Benson et al. [9]; (iii) Tsunoda et al. [7].

DNA, such as the portion that carries the information for ribosomal RNA, have differentiated much more slowly during evolution than have other regions that presumably code for proteins. The studies with DNA lead to the same conclusion as the studies with proteins in estimating the rate of evolution. This rate for an entire species corresponds to 1 to 2 base pair substitutions per year as taking place during evolution in the total genetic DNA of a higher organism containing about 3 billion base pairs. These changes are beneficial or neutral rather than deleterious. The rate of evolutionary change is a phenomenon separate from the rate of mutations. Evolution cannot proceed unless mutations take place, but evolution proceeds as a result of natural selection acting upon the pool of genetic variations which is distributed throughout the millions of individuals in a species.

Let us now consider further aspects of DNA and its role in evolution. A present viewpoint is that randomly-distributed base replacements or point mutations occur incessantly in DNA whenever it is replicated, which takes place whenever cells divide during the life of living organisms. This theory predicts that if two groups of organisms of the same species are separated by geographic barriers, the genetic DNA of one group will slowly and inexorably differentiate from that of the other group until a point is reached at which the descendants of the two original lines of inheritance are so different from each other that they can no longer interbreed. Such a barrier will be produced when a base difference of about 2% is reached [20].

Proteins derive their amino acid sequences directly from the base sequences in messenger RNA, and therefore changes in amino acid sequences will be produced by the changes in the base sequences of DNA. It follows that homologous proteins in two different species will have different amino acid sequences and that the quantitative differences will be proportional to the time of evolutionary separation of the two species and to the biochemical characteristics of the protein; for certain proteins are inherently more flexible in their composition than others.

The theory predicts that any molecule which is directly transcribed from DNA, such as transfer RNA or ribosomal RNA, will undergo differentiation in different species of organisms, and the experimental results so far obtained are in agreement with this.

Evolution of the code

It is possible to apply this model to the genetic code itself. In so doing, we must take consideration of the various mechanisms involved in the translation of the code. Messenger RNA is an exact complementary copy of the specifications in DNA for amino acid sequences in polypeptides, and, as such, has a comparatively passive role to play. The other components of the translation mechanism, such as transfer RNA, are likely to play a more active role in evolution.

An examination of the code (table 1) shows that certain amino acids have more codons than others. Obviously, the fewer codons per amino acid, the more amino acids will be concerned in protein synthesis and the more complex will be the biological system. It therefore seems indicated that, if complex systems evolved from simple ones, there must have been a time when fewer than 20 amino acids were involved in protein synthesis [21].

The most likely candidates for the amino acids used in this earlier era are those which have four or more codons apiece. These are alanine, lysine, valine, threonine, leucine, proline, arginine and serine. We have therefore proposed that at an earlier stage of evolution no amino acid had fewer than four codons. This proposal leads to the concept that a new amino acid is introduced or was introduced into protein synthesis by capturing one or two codons from a group of four that coded for the same amino acid. It is necessary to assume that such changes in the code took place before evolutionary divergence started; in other words, that they took place in the single organism that was, according to current evolutionary hypothesis, the ancestor of all the millions of living forms now present on earth, because in all these forms the code is thought to be universal.

The four codons starting with AU perhaps give a clue as to how an increase took place in the number of amino acids used in protein synthesis. Let us suggest that at one time AUG was a codon for isoleucine. It is next necessary also to assume that the anticodon UAU disappeared. This could be brought about by a mutation that destroyed the function of the tRNA carrying this anticodon. The non-functional gene might then vanish. The codon AUG could change its assignment to methionine if the transfer RNA carrying the anticodon CAU had its properties altered by mutation so that it combined with methionine rather than isoleucine. Methionine would then replace isoleucine throughout all proteins in the organism when the codon AUG was translated [22].

The next step is that one of the codons loses its function in peptide synthesis This is illustrated by the example of the four codons starting with UG. Two of these, UGC and UGU, are codes for cysteine; UGG is the codon for trytophan; and UGA is unassigned to an amino acid but is an interval or chain terminating codon. The stage is now set for the capture of the unassigned codon by the amino acid that has only one codon. This could happen by duplication of the gene for the tRNA which has the anticodon CCA, followed by a point mutation of CCA to UCA. A mutation in the anticodon, GUA to CUA, is known to exist in the anticodon of *E. coli* tyrosine suppressor RNA (table 10).

The result of such a postulated event is illustrated in the next example; that of lysine and asparagine. In this, and in similar examples, a "quartet" of codons is divided equally between two amino acids. Notice that the division is always in the same way; there are no cases where one amino acid is coded by XYA and XYU and the other by XYC and XYG. This pattern fits the wobble hypothesis. All the tRNA molecules can be arranged in the "clover leaf" pattern of secondary structure that was proposed by Holley and coworkers [23]. The general shape of this is shown in fig. 1. It contains the following regions:

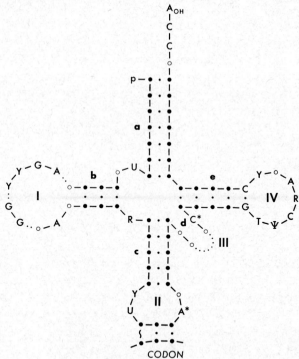

Fig. 1. General formula for transfer RNA molecules, showing helices, loops, and connecting regions. Invariant bases are lettered. ● = hydrogen-bonded bases; ○ = unbonded bases; * designates a base that is sometimes modified; H_I–H_{IV}, helical regions. A messenger codon is shown paired with the anticodon; $\sim$, wobble pairing.

i. Four helical regions, characteristically with seven, four, five and five Watson–Crick base pairs respectively. The first helical region starts with the left-hand terminal nucleotide which has a phosphate group esterified to the 5'-position. Unpaired bases often occur in the helical regions, especially in the first one. It was suggested that these may result from single-base changes occurring during evolution and that they represent the intermediate stage of a change from one base pair to another, e.g., $G=C \rightarrow G \sim U \rightarrow A=U$ [24]. The helical regions are all shown in table 9.

It is suggested that the helicity of these regions is their only essential property, because none of the bases in these regions are identical at corresponding loci in all the known tRNAs except the GC pair which ends the fourth helix and adjoins the $T\Psi C$ loop.

ii. A fifth helical region connects the third and fourth helices in yeast and rat serine, and *E. coli* phenylalanine tRNAs. This may have resulted from crossing-over and recombination. There is evidence of this when the "extra" sections present in yeast serine tRNA and *E. coli* tyrosine tRNA are compared. The comparison is shown in fig. 2. A polynucleotide sequence of 11 residues has apparently been duplicated at a preceding sequence in these two tRNAs, followed by differentiation. The repetition is best seen when yeast tyrosine tRNA, which does not contain the "extra" segment, is compared with *E. coli* tyrosine tRNA.

iii. The "dihydrouridine loop" is variable in size. It contains $-A-G-$; $-G-G-$, and $-A-$ in homologous locations.

iv. The "anticodon loop" contains seven bases. The anticodon is preceded by U and followed by A or modified A.

v. The loop containing the $T\Psi C$ sequence evidently has no amino-acid specificity, for two pairs of tRNAS, valine and phenylalanine yeast tRNAs; and formyl methionine *E. coli* and yeast serine-2 tRNAs, are identical with each other in this sequence of seven nucleotides.

vi. The initial pG is complemented by the terminal base of the fourth helix except in the case of formyl methionine tRNA and it was suggested that this exception might have a functional significance [25]. The dihydrouridine loop of tRNA has been suggested as carrying the recognition site for the amino acid activating enzyme [26]. These two loops are very different for yeast and *E. coli* tyrosine tRNAs as follows:

Yeast $-A-G-hU-hU-Gm-G-hU-hU-hU-A-$

E. coli $-A-G-$ $-C-Gm-G-C-C-A-A-$

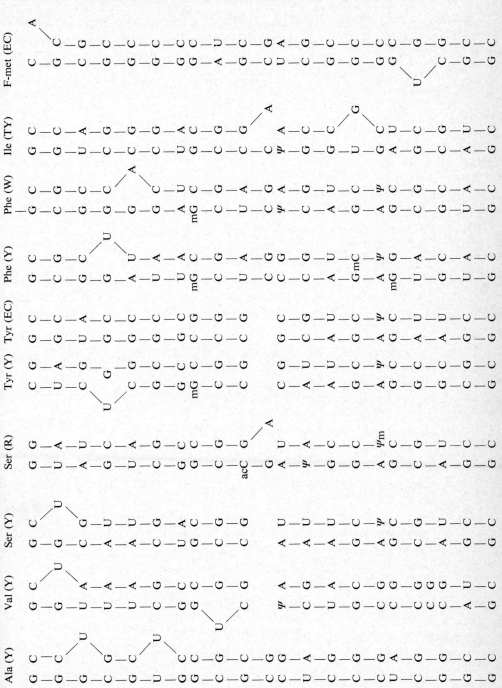

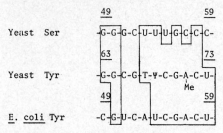

Fig. 2.   Repetitive sequence in Ser and Tyr tRNAs.

but the two anticodon loops are quite similar, as follows:

$$\text{Yeast} \quad -C-U-G-\Psi-A-m_2A-A-$$

$$E.\ coli \quad -C-U-G-U-A-A-\dagger-A-$$

Yeast and *E. coli* tyrosine tRNAs are therefore potentially useful for comparing the effects of the tyrosine tRNA activating enzyme from one of the respective organisms on the tRNA from the other organism. This has been done by Doctor et al. [27]. They found that there was absolute species-specificity. The *E. coli* enzyme would not charge yeast tyrosine tRNA 1 and, vice versa, the yeast enzyme would not charge either of two tyrosine tRNAs obtained from *E. coli* as follows:

Source of tRNA	Source of enzymes	
	E. coli	Yeast
Tyr 1 Yeast	–	1.37
Tyr 1 *E. coli*	1.08	–
Tyr 2 *E. coli*	1.33	–

The figures denote the mμmoles of tyrosine incorporated per unit of tRNA calculated from ultraviolet absorption.

In contrast, when a similar experiment was carried out with yeast and wheat germ phenylalanine tRNAs by Dudock et al. [26], they found that either enzyme would charge either tRNA. The dihydrouridine loops of these two tRNA are identical, as follows:

$$-A-G-hU-hU-G-G-G-A-$$

but the other regions of the two molecules differ from each other. These comparisons led to the conclusion by Dudock et al. [26] that the dihydrouridine loop identifies the tRNA to the activating enzyme. An alternative proposal was made by Schulman and Chambers [28]. They studied the inactivation of alanine tRNA by ultraviolet light, and they concluded that the target for inactivation was located in the first three nucleotide pairs of the top helix (table 9). These are identical in yeast and wheat phenylalanine tRNAs. Perhaps the activating enzyme binds with *both* the dihydrouridine loop and the first three base pairs of the top helix. The comparisons also raise questions of the manner of differentiation of the tyrosine tRNA of yeast from that of *E. coli*. If these two tRNAs had a common origin, how is it possible that they have diverged so much? Each evolutionary change in the recognition site of each tRNA would have to be accompanied by a corresponding adjustment in the primary structure of the activating enzyme. An alternative mechanism would be convergent evolution. The two possibilities are considered below.

The sequences of ten tRNAs, aligned for homology, are in table 10. Fifteen loci are genetically identical in all ten molecules, assuming that the unidentified nucleosides in position 42 are adenine derivatives and are in the unmodi-

TABLE 10

Base sequences of yeast alanine, *E. coli* formylmethionine, yeast and *Torulopsis* isoleucine and valine, yeast and rat serine, yeast and *E. coli* tyrosine, and yeast and wheat phenylalanine transfer RNAs.

	Ala	F	Val	Ile	Ser		Tyr		Phe	
	Y	Met EC	Y	TY	Y	R	Y	EC	Y	W
1	pG	pC	pG	pG	pG	pG	pG	pC	pG	pG
	G	G	G	G	G	U	U	G	C	C
	G	C	U	U	C	A	C	U	G	G
	C	G	U	C	A	G	U	G	G	G
	G	G	U	C	A	U	C	G	A	G
	U	G	C	C	C	C	G	G	U	G
	G	G	G	U	U	G	G	G	U	A
	*U	U†	U	U	U	U	U	Ut	U	U
	mG	G	mG	G	G	G	A	Ut	A	A
10	G	G	G	G	G	G	mG	C	mC	mG
	C	A	U	C	C	C	C	C	C	C
	G	G	C	C	acC	acC	C	C	U	U
	C	C	Ψ	C	G	G	A	G	C	C
	G	–	–	–	–	–	–	–	–	–
	U	–	–	–	–	–	–	–	–	–
	*A	A	A	A	A	A	A	A	A	A

Table 10 (Continued)

	Ala	F	Val	Ile	Ser		Tyr		Phe	
	Y	Met	Y	TY	Y	R	Y	EC	Y	W
		EC								
	*G	G	G	G	G	G	G	G	G	G
	nU	C	hU	hU	hU	hU	hU	–	hU	hU
	C	C	C(hU)	hU	–	–	hU	C	hU	hU
20	–	U	–	–	–	–	–	–	–	–
	*G	G	G	G	Gm	Gm	Gm	Gm	G	G
	*G	G	G	G	G	G	G	G	G	G
	–	–	hU	hU	hU	hU	hU	C	–	–
	–	–	–	–	–	–	hU	C	–	–
	hU	hU	hU(C)	hU	hU	hU	hU	A	G	G
	*A	A	A	A	A	A	A	A	A	A
	G	G	U	A	A	A	A	A	G	G
	C	C	G	G	G	G	G	G	A	A
	G	U	G	G	G	G	G	G	G	G
30	C	C	C	C	C	C	C	G	C	C
	m₂G	G	A	m₂G	m₂G	m₂G	m₂G	A	m₂G	m₂G
32	C	U	Ψ	Ψ	A	A	C	G	C	Ψ
	U	C	C	G	A	Ψ	A	C	C	C
	C	G	U	G	A	G	A	A	A	A
	C	G	G	U	G	G	G	G	G	G
	C	G	C	G	A	A	A	A	A	A
	U	Cm	Ψ	C	Ψ	mC	C	C	Cm	Cm
	*U	U	U	U	U	U	U	U	U	U
	I	C	I	I	I	I	G	G†(C)	Gm	Gm
40	G	A	A	A	G	G	Ψ	U	A	A
	C	U	C	U	A	A	A	A	A	A
	*mI	A	A†	A†	iA	iA	iA	A†	N	N
	Ψ	A	C	A	A	A	A	A	A	A
	G	C	G	C	Ψ	Ψm	Ψ	Ψ	Ψ	Ψ
	G	C	C	G	C	C	C	C	mC	C
	G	C	A	C	U	C	U	U	U	U
	A	G	G	C	U	A	U	G	G	G
	G	A	A	A	U	U	G	C	G	A
	A	A	A	A	Um	Um	A	–	A	A
50	G	G	C	G	G	G	G	C	G	G
	–	mG(A)	–	A	G	G	A	G	mG	N
	–	–	–	–	G	G	–	U	–	–
	–	–	–	–	C	G	–	C	–	–
	–	–	–	–	U	U	–	A	–	–
	–	–	–	–	U(C)	mC	–	C(U)	–	–
	–	–	–	–	U	U	–	A(C)	–	–
	–	–	–	–	G	C	–	G	–	–
	–	–	–	–	C	C	–	A	–	–
	–	–	–	–	C	C	–	C	–	–
60	–	–	–	–	C	C	–	U	–	–
	U†	U	hU(−)	hU	G	G	hU	U	U	C

Table 10 (Continued)

	Ala Y	F Met EC	Val Y	Ile TY	Ser Y	Ser R	Tyr Y	Tyr EC	Phe Y	Phe W
	*C	C	mC	mC	mC	mC	mC	C	C	C
63	U	G	C	A	G	G	G	G	mC	G
	C	U	C	G	C	C	G	A	U	C
	C	C	C	C	A	A	G	A	G	G
	G	G	A	A	G	G	C	G	U	U
	*G	G	G	G	G	G	G	G	G	G
	*T	T	T	T	T	T	T	T	T	T
	*ψ	ψ	ψ	ψ	ψ	ψ	ψ	ψ	ψ	ψ
70	*C	C	C	C	C	C	C	C	C	C
	G	A	G	G	G(A)	G	G	G	G	G
	*A	A	mA	mA	A	mA	mA	A	mA	mA
	U	A	U	U	G(A)	A	C	A	U	U
	U	U	C	C	U	U	U	U	C	C
	*C	C	C	C	C	C	C	C	C	C
	C	C	U	U	C	C	G	C	A	A
	G	G	G	G	U	U	C	U	C	C
	G	G	G	C	G	G	C	U	A	G
	A	C	G	U	C	C	C	C	G	C
80	–	–	G(–)	–	–	–	–	–	–	–
	C	C	C	A	A	C	C	C	A	U
	U	C	G	G	G	G	C	C	A	C
	C	C	A	G	U	A	G	C	U	A
	G	C	A	G	U	C	G	C	U	C
	U	G	A	A	G	U	G	A	C	C
	C	C	U	C	U	A	A	C	G	G
	C	A	C	C	C	C	G	C	C	C
88	A	A	A(–)	A	G	G	A	A(C)	A	A

C
C
A-OH

Abbreviations: Y, yeast; EC, *E. coli*; W, wheat; N, unspecified nucleoside; hU, dihydrouridine; Ut, 4-thiouridine; Cm, O-methyl cytidine; Gm, O-methyl guanosine; mG, methyl guanosine; m^7G, 7-methyl guanosine; mA, methyladenosine; ψ, pseudouridine; m_2G, dimethylguanosine; mI, methyl inosine; m_2A, dimethyl adenosine; m_2C, dimethyl cytidine; iA, isopentenyl adenine; acC, acetylcytidine; mC, methylcytidine; †, modified nucleoside; *, designates nucleosides that are identical in all the tRNAs listed; lines separate the helical regions; hyphens indicate evolutionary "gaps" [10]. Letters in parentheses are residues in variants of the tRNAs beside which they appear; those beside the yeast valine sequence are differences that are present in *Torulopsis utilis* valine tRNA.

Sources: Holley et al. [23]; Zachau et al. [19]; Madison et al. [30]; RajBhandary et al. [31]; Bayev et al. [32]; Takemura et al. [33]; Goodman et al. [34]; Dube et al. [25]; Dudock [35]; Nishimura et al. [36]; Staehelin et al. [37].

fied form at the time of transcription. The random probability for identity of any two sites in two RNA sequences is about 25%, so that there will be an expectation of about 20 identical sites in two tRNA sequences, but less than one in more than four sequences.

There is a marked homology between yeast and wheat phenylalanine tRNAs and between yeast and rat serine tRNAs (table 10). These tRNAs share with the other tRNAs in table 10, the identity at the DNA level of the fifteen bases marked by *; some of these bases are modified subsequently to transcription by methylation, etc. There are 58 other bases in the two phenylalanine tRNAs in addition to the 15 that are identical in all ten tRNAs. Of these 58, 43 are identical and 15 are different in the yeast; wheat comparison: a difference of 26%. In the analogous comparison between yeast and rat serine tRNAs, there are 67 bases in addition to the "invariable" 15. Of these 67, 49 are identical and 18 are different; a difference of 25%. These results may be compared with the differences between the cytochromes c of baker's yeast, wheat and a mammal (rabbit). Rabbit is used as an example of a mammal because the primary structure of rat cytochrome c has not been reported. The yeast:wheat difference is 43 amino acid residues in 107 sites compared, corresponding to 61 minimum base differences in 107 codons, and the yeast:rabbit difference is 39 amino acid residues in 103 sites, corresponding to 55 minimum base differences, in 103 codons. If it is assumed that 29 of the amino acids are invariant, the 61 minimum base differences between yeast and wheat cytochromes c in the remaining 78 codons are equivalent to 0.78 MBDC or 26 base substitutions per 100 base pairs. In the yeast:rabbit comparison, there are 55 minimum base differences between $103 - 29 = 74$ codons, equivalent to 0.74 MBDC or 25 base substitutions per 100 base pairs. The minimum base differences presumably correspond to a larger number of actual base differences because of ambiguity in the third position of codons. The conclusion may be drawn that the cistrons for the phenylalanine tRNAs of yeast and wheat and for the serine tRNA of yeast and mammals have diverged during evolution at about the same rate or perhaps somewhat more slowly than have the cistrons for the cytochromes c of these organisms. The comparisons of the tRNAs and the cytochromes c show that the phanerogams, ascomycetes and mammals may have differentiated to approximately equal extents from a common ancestor in terms of evolutionary replacement of DNA base pairs. It is impressive that the same results are obtained in the tRNA comparisons and the cytochrome c comparisons.

The clear-cut nature of the evolutionary divergence of these two phenylalanine tRNAs and two serine tRNAs encourages an attempt to compare the two tyrosine tRNAs (table 11). Without the availability of the yeast:wheat and yeast:mammalian comparisons, one might conclude that it was difficult to represent the large difference between the two tyrosine tRNAs in terms of divergent evolution. This difference is 32 bases in 58 comparisons, or 55% which is about twice that of the yeast:wheat and yeast:mammalian comparisons. This infers that the evolutionary separation of bacteria from the common ancestor of higher plants, vertebrates and yeasts took place about twice as long ago as the yeast: vertebrate higher plant separation if it is assumed that the rate of differentiation in the tRNAs is approximately linear with time.

The next comparison that can logically be made is that of a tRNA for one amino acid with a tRNA of a different amino acid. This should have a bearing on the origin of the amino acid code. The results are in table 11. The average of all such comparisons, excluding, of course, the yeast: *E. coli* tyrosine, yeast:rat serine, and yeast:wheat phenylalanine comparisons, is 60.9%. The difference is not much greater than the yeast: *E. coli* tyrosine difference of 55%, a finding which raises the interesting possibility that the genetic code originated not long before the evolutionary separation of the lines of descent leading to yeast and *E. coli* took place. There is a considerable amount of variation in the comparisons between tRNAs for two different amino acids;

TABLE 11

Percent differences between homologous sites of eight transfer RNA molecules, excluding sites common to all eight

	Y Ala	EC F Met	Y Val	TY Ile	Y Ser	R Ser	Y Tyr	EC Tyr	Y Phe	W Phe
Y Ala		56	54	56	63	57	69	70	65	67
EC F Met	56		65	58	68	60	64	56	69	62
Y Val	54	65		47	65	61	69	63	68	63
TY Ile	56	58	47		61	62	58	72	62	64
Y Ser	63	68	65	61		(25)	52	57	58	62
R Ser	57	60	61	62	(25)		54	52	67	57
Y Tyr	69	64	69	58	52	54		(55)	53	52
EC Tyr	70	56	63	72	57	52	(55)		64	55
Y Phe	65	69	68	62	58	67	53	64		(26)
W Phe	67	62	63	64	62	57	62	55	(26)	
Average of comparisons between pairs of different amino acids	61.9	62.0	61.7	60.0	60.8	58.8	58.9	61.1	63.2	60.2

the greatest difference is between *Torulopsis* yeast isoleucine and *E. coli* tyrosine tRNAs. The difference between *Torulopsis utilis* valine and isoleucine tRNAs is the smallest of all comparisons of tRNAs for different amino acids. This is of interest in view of the chemical similarity between valine and isoleucine. The average of the differences between pairs of tRNAs that are related by a single base change in anticodons is 58.6%. The corresponding figure for tRNAs in which the anticodons differ by two base changes is higher, 63.9%.

Another possibility to be considered is that the tRNAs for different amino acids have been separated so long that they have reached an equilibrium. This would represent a random difference of about 75%, for there is about one chance in four for two of the four bases to be identical at random. The values in table 11 are for the most part substantially less than 75%. This suggests that evolution in the tRNA series has not reached an equilibrium with respect to the differences between tRNAs for different amino acids.

Missing from table 11 is a comparison between two tRNAs from the same organism in which the anticodons are different, for example, yeast alanine I, IGC anticodon and yeast alanine II, UGC anticodon. Such a comparison might indicate the time that has elapsed since the code has changed from a "two-letter code" (i.e., one in which a single anticodon paired with four codons) to the present code. It is also possible, however, that a tRNA containing a certain anticodon might disappear and reappear by gene duplication and differentiation, since only one base change is necessary to change, e.g., one alanine to any other alanine anticodon, and the remainder of the molecule could continue to function in the acceptance and transfer of alanine after the change in the anticodon.

We assume that the evolution of the code ceased at the stage shown in table 1. If the evolution had continued to increase the number of amino acids in protein synthesis, a biological system could have developed in which 30 or 31 amino acids participated in protein synthesis. To proceed even further into diversification requires that the wobble pairing be replaced by something more specific. A system would then be possible where 61 amino acids, each with one codon, participate in protein synthesis. For this, it would be necessary to have further differentiation of the transfer RNAs and to have a biological system sufficiently flexible so that new amino acids could be introduced into proteins.

But this did not happen. Instead the code froze in its present form as soon as divergent evolution started. From this point on, the complexity of the biochemistry of terrestrial life was too great to permit changes throughout the proteins of any organism. The complexity increased with time, so that introduction of a new amino acid into protein synthesis would undoubtedly be lethal.

The model of evolution that is presented leads back to a single organism which used the present code. This organism had competitive advantages so great that its descendants have crowded out all other forms of life. They have done this by divergent evolution. Prior to this ancestral organism, we conjecture there was a long period of parallel or convergent evolution in which more primitive codes and proteins existed.

Summary

Evolution depends upon the occurrence of occasional changes, large or small, in hereditary characteristics. Molecular genetics gave rise to the new field of molecular evolution, which is currently exploring the changes that take place in proteins and nucleic acids over long periods of time. The following are some of the fundamental assumptions:

1. The phenotypic characteristics of organisms depend directly on proteins.
2. Proteins are synthesized in accordance with information carried in molecules of DNA as sequences of the four bases, adenine, guanine, cytosine, and thymine. The information is transcribed into molecules of messenger RNA and is translated into proteins by the intervention of the genetic code.
3. Changes in the composition of the base sequences in DNA can take place in living organisms, and these changes can affect the phenotypic characteristics of the next generation.
4. The process of natural selection favors the perpetuation of organisms which compete successfully in the struggle for existence. This process leads to the elimination of all but a small fraction of the astronomical number of possible protein molecules that could result from genetic translation of the possible variants of DNA. Furthermore, the number of protein molecules was originally much smaller than it is to-day, and it has increased by hereditary processes rather than by the chance appearance of entirely new proteins.
5. The DNA present in any single cell contains the complete information for all the hereditary characteristics of the organism. The amount of DNA per cell may increase during evolution and this increase has produced modern organisms that are "higher", more specialized, and more complex, from earlier and simpler forms.

6. Protein molecules are slowly and steadily differentiated during evolution if their genes are physically separated from each other, by allopatric speciation or even by duplication and translocation, whether or not the functions of the proteins are changed.

7. Mutations, together with recombination, contribute to changes in the genetic pool which provide the variability within populations that is necessary for evolution of species.

The field of molecular evolution should include a theory of the chemical events leading to the formation of the first living organism from molecules of non-living origin.

The genetic code may have evolved through multiplication of transfer RNA molecules by gene duplication followed by differentiation. This proposal is supported by the similarities between all tRNA molecules of known structures.

The DNA of higher organisms contains families of repetitive sequences. The families may contain thousands or hundreds of thousands of individual members. The "family resemblance" within each group grows less with the passage of time because this leads to differentiation resulting from the accumulation of point mutations.

Acknowledgment. The author thanks Dr. C. R. Cantor for numerous discussions and suggestions.

References

[1] J. D. Watson and F. H. Crick, Nature *171* (1953) 737.

[2] W. C. Summers and W. Szybalski, Virology *34* (1968) 9.

[3] L. Pauling and R. B. Corey, Arch. Biochem. Biophys. *65* (1956) 164.

[4] F. H. C. Crick, J. Mol. Biol. *19* (1966) 548.

[5] H. Matsubara, R. M. Sasaki and R. K. Chain, Proc. Natl. Acad. Sci. U.S. *57* (1967) 439.

[6] H. Matsubara and K. Sugeno, (1968) in preparation.

[7] J. Tsunoda, H. Whiteley and K. T. Yasunobu, Pacific Slope Biochemical Conference, abstracts, (1967) p. 103.

[8] M. Tanaka, T. Nakashima, A. Benson, H. F. Nower and K. T. Yasunobu, Biochem. Biophys. Res. Commun. *16* (1964) 422.

[9] A. M. Benson, H. F. Mower and K. T. Yasunobu, Proc. Natl. Acad. Sci. U.S. *55* (1966) 1532.

[10] C. R. Cantor, Biochem. Biophys. Res. Commun. *31* (1968) 410.

[11] J. Marmur and D. Lane, Proc. Natl. Acad. Sci. U.S. *46* (1960) 453.

[12] C. L. Schildkraut, K. L. Wierzchsowski, J. Marmur, D. M. Green and P. Doty, Virology *18* (1962) 43.

[13] E. T. Bolton and B. J. McCarthy, Proc. Natl. Acad. Sci. U.S. *48* (1962) 1390.

[14] B. H. Hoyer, B. J. McCarthy and E. T. Bolton, Science *140* (1963) 1408.

[15] B. H. Hoyer, B. J. McCarthy and E. T. Bolton, Science *144* (1964) 959.

[16] B. J. McCarthy and E. T. Bolton, Proc. Natl. Acad. Sci. U.S. 50 (1963) 156.
[17] B. J. McCarthy and E. T. Bolton J. Mol. Biol. 8 (1964) 184.
[18] B. J. McCarthy, and B. H. Hoyer, Proc. Natl. Acad. Sci. U.S. 52 (1964) 915.
[19] R. J. Britten and D. E. Kohne, Carnegie Institution of Washington Year Book 65 (1965-6) p. 78.
[20] B. J. McCarthy, Presented at the Western Experiment Station Collaborators Conference; Albany, California (March 13, 1968).
[21] T. H. Jukes, Molecules and evolution (Columbia Univ. Press, New York, 1966).
[22] T. H. Jukes, Biochem. Biophys. Res. Commun. 27 (1967) 573.
[23] R. W. Holley, J. Apgar, G. A. Everett, J. T. Madison, M. Marquisee, S. H. Merrill, J. R. Penswick and A. Zamir, Science 147 (1965) 1462.
[24] T. H. Jukes, Biochem. Biophys. Res. Commun. 24 (1966) 744.
[25] S. K. Dube, K. A. Marcker, B. F. C. Clark and S. Cory, Nature 218 (1968) 232.
[26] B. A. Dudock, G. Katz, E. K. Taylor and R. W. Holley, Federation Proc. 27 (1968) 342.
[27] Doctor et al. (1966)
[28] L. H. Schulman and R. W. Chambers, presented at Symposium on Transfer RNA, New York, (June 7-8, 1968).
[29] H. G. Zachau, D. Dütting and H. Feldmann, Hoppe-Seylers Z. Physiol. Chem. 347 (1966) 212.
[30] J. T. Madison, G. A. Everett and H. K. Kung, Science 153 (1966) 531.
[31] U. L. RajBhandary, S. H. Chang, A. Stuart, R. D. Faulkner, R. M. Foskinson and H. G. Khorana, Proc. Natl. Acad. Sci. U.S. 57 (1967) 751.
[32] A. A. Bayev, T. V. Venkstern, A. D. Mirzabekow, A. I. Krutilina, L. Li and V. D. Axelrod, Mol. Biol. (USSR) 1 (1967) 754.
[33] S. Takemura, T. Mizutani and M. Miyazaki, J. Biochem. 63 (1968) 277.
[34] H. M. Goodman, J. Abelson, A. Landy, S. Brenner and J. D. Smith, Nature 217 (1968) 1019.
[35] B. A. Dudock, presented at Symposium on Transfer RNA, New York, (June 7-8, 1968).
[36] S. Nishimura and co-workers, (1968) personal communication.
[37] M. Staehelin, H. Rogg, B. C. Baguley, T. Ginsberg and W. Wehrli, Nature 219 (1968) 1363.

A CONTRIBUTION TO THE EVOLUTION
OF STRUCTURAL PROTEINS

A. NORDWIG and U. HAYDUK

*Max-Planck-Institut für Eiweiss- und Lederforschung, Abt. Kühn,
München, Germany*

This report does not deal with the whole amino acid sequence of a protein. However, certain characteristics of collagen, a fibrous protein occurring in any multicellular animal organism as the main component of connective tissue, enabled us to draw some common conclusions with regard to the phylogenetic fate of this protein.

One reason for the fact that the primary structure of collagen is known only partially is its high molecular weight of about 280,000 [1]. The molecule is made up of three subunits of equal size, the so-called α-components, each containing approximately 1030 amino acid residues. The tertiary structure of the collagen molecule is a triple helix of polyproline II type with the three α-polypeptide chains intertwined about a common axis.

The arrangement of collagen molecules in the naturally occurring fiber is shown in fig. 1, where each arrow stands for a triple-stranded molecule. This quarter-staggered array gives rise to a cross-striation visible in the electron-microscope, the period of this pattern being approximately 650 Å. Several groups have presented convincing evidence [1] that the light bands of the electron-microscopic pattern correspond to the so-called apolar regions of the molecule which are tripeptide polymers structured $(Gly-Pro-X)_n$. The dark bands could be correlated to the polar sequence regions consisting predominantly of clusters of basic or acidic amino acids of unknown sequence.

In vitro one can produce from collagen solutions aggregates where the molecules are arranged with their ends in register (fig. 2). Thus, these artificial quaternary structures called segments long-spacing (SLS, 2800 Å in length) allow direct observation of the sequence regions mentioned above without any overlap as in the case of fibers (fig. 1). They are, therefore, the basis for gaining some insight into the gross array of the collagen primary structure.

We have isolated and purified the acid-soluble collagens from *Actinia equina*,

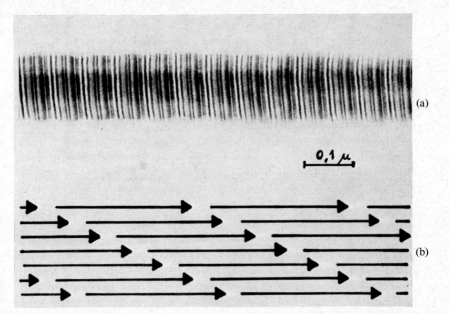

Fig. 1. (a) Electronmicrograph of a calf skin collagen fiber, positively stained with phosphotungstic acid and uranylacetate. (b) Schematic representation of molecule array in a collagen fiber (cf. text). (Taken from B. Zimmerman, Thesis, Univ. of Heidelberg, 1966. Micrograph taken by K. Kühn.)

a kind of sea anemone, and from the liver fluke *Fasciola hepatica*, a parasite living in the bile ducts of sheep and cattle [2]. These invertebrate proteins exhibited essentially the well-known chemical and physico-chemical properties of vertebrate material (collagens from calf skin and carp swim bladder were used for comparison). There were some significant deviations in the amino acid compositions, however [2]. It is true, the striking characteristics of amino acid analyses of vertebrate collagens were also observed here: one third of the residues were glycine and high amounts of imino acids were present as well as the two unusual amino acids, hydroxyproline and hydroxylysine. Quantitatively, the contents of proline, 4-hydroxyproline, alanine, methionine and hydroxylysine differed significantly from collagen to collagen. In addition, 3-hydroxyproline occurred in collagen from *Actinia* in appreciable amounts. These findings suggested non-identity of amino acid sequence of collagens derived from phylogenetically different species.

Very surprisingly, these differences in amino acid composition do not give

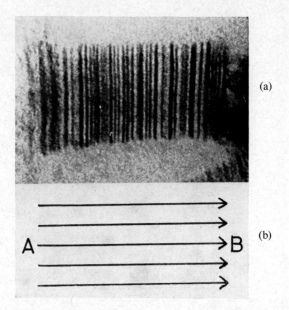

Fig. 2.   (a) Electronmicrograph of a segment long-spacing (SLS) of calf skin collagen. Positive staining with uranyl acetate. (b) Schematic representation of molecule array in a collagen segment (cf. text). A, N-terminus; B, C-terminus of the molecules. (Taken from B. Zimmermann, Thesis, Univ. of Heidelberg, 1966. Micrograph taken by K. Kühn.)

rise to disturbance of triple helix or quaternary structure formation. When SLS of the various collagens were investigated in the electron-microscope, extraordinary similarity of the structures with respect to length as well as to the cross-striation pattern of the segments were observed (fig. 3). As the correct triple helical tertiary structure is a prerequisite for the correct quaternary structures, one may conclude that both spatial structures of collagen have kept constant during phylogenetic development [2, 3].

One has to consider, however, the differences in primary structure. This apparently means that not the amino acid sequence as a whole contains the information for helix and fibril formation. This information seems to be coded in certain features of the primary structure, and it is these features that are not subject to changes by mutation. There is little doubt that these features are the content of one third of glycine, important for the correct tertiary structure to be formed, and the regular alteration of polar and apolar sequence regions mentioned before. The latter property provides the basis

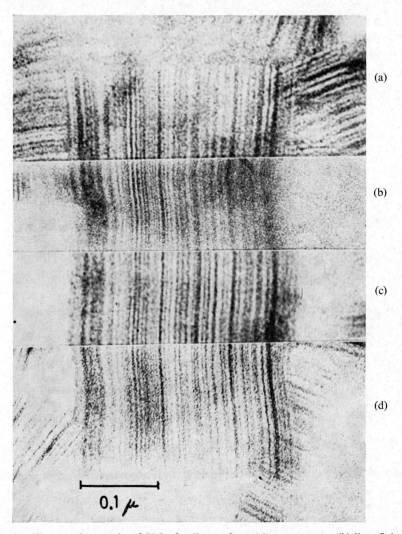

Fig. 3. Electronmicrographs of SLS of collagens from (a) sea anemone, (b) liver fluke, (c) carp swim bladder, (d) calf skin. Positive staining as in fig. 1. Number, relative positions and relative intensities of the bands of the individual segments are practically identical (cf. text). (Taken from [3].)

for a certain charge profile along the collagen molecule and this charge profile in turn is responsible for the correct quaternary structures. We have been able to confirm this idea by comparing the cross-striation patterns of collagen fibrils from calf skin and *Fasciola hepatica* [2].

Thus, several concepts developed for the evolution of globular proteins appear to be valid also for collagen as a representative of fibrous material: homology of subunit polypeptide chains and constancy of configuration as the limiting factor for evolutionary processes. To the best of our knowledge, the present work is the first contribution to the chemical evolution of structural animal proteins.

There are two more points to be made: (1) After denaturation by heat or urea of collagen from *Actinia* we only find one band in the α-subunit region, whereas two or three show up with vertebrate material. This supports the view [4] that only one primordial gene originally existed for the α-polypeptide chain of collagen. Obviously, duplication of the gene then occurred on a higher phylogenetic stage thus explaining the existence of two or three α-chains per molecule of slightly different sequence in vertebrate collagens. (2) When measuring the stability of the various collagens in solution (by following the optical rotation when the temperature is slowly elevated) we have found a melting temperature of 26.6 °C for the *Actinia* collagen and of 36.6 °C for the *Fasciola* protein (calf skin collagen, 39 °C). This clearly indicates that thermal stability of collagen is not a function of evolution but is rather correlated to the habitat conditions of the animal from which the protein was extracted.

We are very grateful to Mrs. H. Wiedemann, Mrs. I. Scholz and Prof. K. Kühn for their help with respect to electron-microscopy.

References

[1] For reviews on collagen chemistry see W. Graßmann et al., Fortschr. Chem. Org. Naturstoffe *23* (1965) 195, and Treatise on collagen, G. N. Ramachandran, ed., Vol. I (Academic Press, New York, 1967).
[2] A. Nordwig and U. Hayduk, J. Mol. Biol. *44* (1969) 161.
[3] A. Nordwig and U. Hayduk, J. Mol. Biol. *26* (1967) 351.
[4] J. Pikkarainen and E. Kulonen, Scand. J. Clin. Lab. Invest., Suppl. *95* (1967) 40; J. Pikkarainen, Acta Physiol. Scand., Suppl. *309* (1968) 1.

CELLULAR DIFFERENCES IN THE TRANSFER RNA CONTENT

C. THOMAS CASKEY

National Heart Institute, Bethesda, Maryland, U.S.A.

It is now possible to assign for *Escherichia coli* the 64 triplet mRNA codons to one of 20 amino acids or peptide chain termination function [1–12]. Comparative studies of mammalian, amphibian, and bacterial amino-acyl-tRNA [13], however, have demonstrated that cellular differences in mRNA codon recognition occur. The regulatory effects of such cellular differences in mRNA codon recognition are well known in the cases of bacterial strains containing nonsense [14–16] and missense [17], extra-genic suppressor genes. The products of these suppressor genes, tRNA [18], as a result of altered codon recognition can correct premature chain termination mutations (UAA, UAG, or UGA) and unacceptable missense mutations by insertion of acceptable amino acids. These and other studies clearly illustrate that biologic components required for the translation of mRNA (t-RNA, aminoacyl-tRNA synthetase, mRNA, ribosomes, and protein synthesis factors) exert a regulatory effect on protein synthesis. The potential for tRNA serving a translational regulatory note has been explored in depth in these studies. The following studies will present evidence supporting these conclusions: (1) the *amino acid code* of widely different cellular types is predominately *universal*; (2) messenger RNA codons differ in their template acitivity for aminoacyl-tRNA from bacterial, mammalian, and amphibian sources; (3) the *transfer RNA code* of mammals and bacteria differ.

The recognition values of mRNA codons by mammalian, and bacterial amino-acyl-tRNA have been directly compared. Following tRNA isolation from the three sources and acylation with [^{14}C]- or [^{3}H]-amino acids by the homologous aminoacyl-tRNA synthetase, codon recognition was determined by triplet mRNA codon directed aminoacyl-tRNA ribosomal binding as described by Nirenberg and Leder [1]. The synonym codon recognition values for 17 unfractionated (liver) aminoacyl-tRNAs from guinea pig were qualitatively identical to the corresponding 17 unfractionated (*E. coli*) aminoacyl-tRNA. Quantitative differences in the *relative template activity of synonym codons* for (liver) and (*E. coli*) aminoacyl-tRNA, do occur, however. Examples of such quantitative differences are shown in table 1. Since the conditions and components for the determination of codon recognition differ only by the source of [^{14}C] or [^{3}H]-aminoacyl-tRNA, all differences in codon re-

TABLE 1

Codon recognition by (*E. coli*) and (liver) aminoacyl-tRNA

Trinucleotide	$\Delta\mu\mu$mole [^{14}C] or [^{3}H] AA-tRNA Bound to ribosomes	
	Guinea pig liver	*E. coli*
Arg-tRNA		
CGU	0.81	0.90
CGC	0.67	0.47
CGA	1.28	1.09
CGG	0.97	0.20
AGA	0.12	0.10
AGG	0.63	0.12
None ($\mu\mu$mole)	(1.23)	(1.27)
Ile-tRNA		
AUU	1.78	0.64
AUC	0.60	0.73
AUA	0.38	0.01
None ($\mu\mu$mole)	(0.81)	(0.15)
Lys-tRNA		
AAA	0.33	1.00
AAG	0.50	0.07
None ($\mu\mu$mole)	(0.69)	(0.70)
Ser-tRNA		
UCU	1.21	1.27
UCC	0.13	0.54
UCA	0.77	1.56
UCG	0.50	1.09
AGU	0.77	0.21
AGC	0.83	0.26
None ($\mu\mu$mole)	(0.68)	(0.43)

The effects of trinucleotides upon the binding of (*E. coli*) and (liver) aminoacyl-tRNA. Each 0.050 ml reaction was incubated 15 min at 24 °C and contained 0.05 M tris-acetate; 0.05 M potassium acetate; 0.02 M magnesium acetate; 2.0 A^{260} units *E. coli* ribosomes; 0.150 ± 0.010 A^{260} units of trinucleotide; and [^{14}C] or [^{3}H]-aminoacyl-tRNA [13]. Ribosomes were washed on Millipore filters [1] and dried, and radioactivity was determined in a liquid-scintillation counter. All assays were performed in duplicate.

cognition reflect cellular differences in tRNA. It is possible therefore to order the template activity of *synonym codons* for each aminoacyl-tRNA preparation. Codon recognition values for each aminoacyl-tRNA are internally controlled since they are determined under identical conditions. Differences in the order of response of (liver) and (*E. coli*) aminoacyl-tRNA to a synonym codon are referred to as *differences in relative template activity of synonym codons* and shown in table 1. Both (*E. coli*) and (liver) Arg-tRNAs recognize best the codon CGA. (*E. coli*) and (liver) Arg-tRNA, however, differ markedly in their recognition of CGG and AGG. Recognition by (liver) Arg-tRNA of CGG and AGG was 76% and 51% the level of response to CGA. Recognition of (*E. coli*) Arg-tRNA to both codons CGG and AGG was $\frac{1}{5}$ this relative response to CGA. Similarly (*E. coli*) and (liver) Ser-tRNA differ in their order of recognition of synonym codons. Less subtle differences in recognition of mRNA codons by (*E. coli*) and (liver) Ile- and Lys-tRNA were observed. While the codon AUA stimulated substantial ribosomal binding of (liver) Ile-tRNA, none was detected with (*E. coli*) Ile-tRNA. The lysine codon AGG stimulated (liver) Lys-tRNA ribosomal binding markedly while (*E. coli*) Lys-tRNA recognized the codon minimally.

A summary of differences in relative template activity of certain codons for (*E. coli*), (*Xenopus, Laevis*) and (guinea pig) aminoacyl-tRNA is given in table 2. Some codons are active templates with aminoacyl-tRNA from every

TABLE 2

Codon		tRNA		
		Bacterial (*E. coli*)	Amphibian (X. *laevis*, liver)	Mammalian (G. pig, liver)
Arg	AGG	±	+ + + +	+ +
	CGG	±	+ + + +	+ + + +
Met	UUG	+ +	±	±
Ala	GCG	+ + + +	±	+ +
Ile	AUA	±	+ +	+ +
Lys	AAG	±	+ + + +	+ + + +
Ser	UCG	+ + + +	±	+ +
	AGU	±	+ + +	+ + +
	AGC	±	+ + +	+ + +

Species-dependent differences in response of aminoacyl-tRNA to trinucleotide codons. The following scale indicates the approximate response of aminoacyl-tRNA to a trinucleotide relative to the responses of the same AA-tRNA preparation to every other trinucleotide for that amino acid: + + + +, 70 to 100 percent; + + +, 50 to 70 percent; + +, 20 to 50 percent; ±, 0 to 20 percent.

organism studied; whereas, other codons are active with aminoacyl-tRNA from one or two, but not from every organism studied. While bacterial aminoacyl-tRNA codon recognitions differ from bacterial and mammalian responses, amphibian and mammalian aminoacyl-tRNA respond similarly to trinucleotide codons. Therefore numerous changes in tRNA have occurred between the appearance of bacteria and vertebrates. The earliest dating of bacteria is approximately 3.1 billion years ago [19]; amphibians 355 million years ago; and mammals 174 million years ago. Thus, on the basis of *codon relative template activity* comparatively little change in synonym codon order of recognition has occurred for 355 million years. These data suggest that synonym codon dominance or usage was established early in vertebrates and has undergone little change.

1st Base	2nd Base				3rd Base
	U	C	A	G	
U	PHE	SER	TYR	CYS	U
	PHE	SER	TYR	CYS	C
	leu	SER	term	term	A
	leu, F-MET	SER	term	TRP	G
C	leu	PRO	HIS	ARG	U
	leu	PRO	HIS	ARG	C
	leu	PRO	gln	ARG	A
	leu	PRO	gln	ARG	G
A	ILE	THR	asn	SER	U
	ILE	THR	asn	SER	C
	ILE	THR	LYS	ARG	A
	MET, F-MET ?	THR	LYS	ARG	G
G	VAL	ALA	ASP	GLY	U
	VAL	ALA	ASP	GLY	C
	VAL	ALA	GLU	GLY	A
	VAL, F-MET ?	ALA	GLU	gly	G

Fig. 1.   Nucleotide sequences of RNA codons recognized by aminoacyl-tRNA from bacteria, amphibian liver and mammalian liver were determined by stimulating, with trinucleotide codons, the binding of AA-tRNA to *E. coli* ribosomes. Capital letters indicate aminoacyl-tRNA from the three sources were assayed with trinucleotides, lower case letters indicate only (*E. coli*) aminoacyl-tRNA was assayed.

The mRNA code as determined for mammals, amphibians and bacteria is summarized in figure 1. It is clear that the mRNA code is universal; i.e., mRNA codons designate the same amino acids in all species studied. Six patterns of degenerate codon recognition are apparent. Messenger RNA codons for a single amino acid which differ in the 3rd base (3'-position) position of the codon may occur in the following equivalent sets: (1) U=C (i.e., phenylalanine); (2) A=G (i.e., glutamic); (3) U=A=C=G (i.e., threonine); and (4) U=C=A (i.e., isoleucine). One example, each of degenerate codon recognition occurring in the 1st (5′-position) and middle position of the mRNA codon: (1) A=G≫U for formyl-methionine the amino acid involved in initiation of *E. coli* protein synthesis; (2) A=G, terminator codons for *E. coli*. Although the middle base position degeneracy for the termination function has not been confirmed at the present time, all other mammalian amino acid mRNA codon degenerate sets are identical to those of bacteria.

Within the framework of the *universal code*, however, certain codons appear to vary in template activity for aminoacyl-tRNA depending on its cellular source. These codons, by their variable template activity, may function as translational regulatory sites effecting either absolute or relative rate of mRNA translation containing the codons. It is for this reason that we have pursued the investigation of those aminoacyl-tRNAs whose synonym codons vary in template activity among species. To clarify the dissimilar responses, liver and *E. coli* AA-tRNA preparations were fractionated by reverse phase column chromatography (RPCC), and the ability of trinucleoside diphosphate codons to direct ribosome binding of purified aminoacyl-tRNA determined [20].

Escherichia coli and guinea pig liver tRNA were acylated with [^{14}C]- or [^{3}H]-amino acids in the presence of the corresponding aminoacyl-tRNA synthetase and reactions deproteinized prior to fractionation. All preparations were fractionated by RPCC as described by Kelmers, Novelli, and Stulberg [21] except that all solutions were adjusted to *p*H 4.5 to minimize deacylation.

The fractionation of (*E. coli*) and (guinea pig liver) [^{14}C]-Arg-tRNA is illustrated in fig. 2. Following elution of the Arg-tRNA with a linear NaCl gradient, the fractions indicated by the cross hatched areas were pooled, desalted, and lyophilized. Trinucleotide directed binding of each Arg-tRNA fraction to *E. coli* ribosomes was determined as given in the upper portion of part A and B of fig. 2. *E. coli* ribosomes were used for all binding studies so that codon recognition could be investigated under uniform conditions. *Escherichia coli* Arg-tRNA was resolved into two peaks. The larger of the two peaks responds to the codons CGU, CGC, and CGA while the smaller peak responds only to CGG. Identical responses were observed with (liver) Arg-tRNA fractions two and four. Three additional (liver) Arg-tRNA species were also found. Fraction one responds preferentially to AGG and fractions three and five contain redundant tRNA species responsive to CGA and CGG. Fractionation of isoaccepting (liver) and (*E. coli*) Arg-tRNA

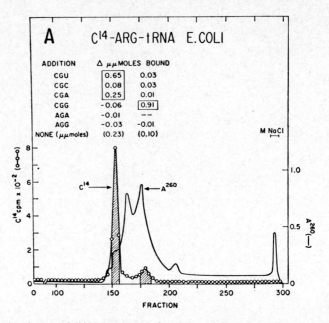

Fig. 2.(a) A mixture of [^{14}C]-Arg-tRNA from *E. coli* B (1.8 × 10^6 cpm precipitated in 5% TCA at 3 °C, 150 A^{260} units) and 360 A^{260} units of carrier *E. coli* tRNA were fractionated by reverse phase column chromatography. (0--0--0) represents total [^{14}C]- or [^{3}H]-counts per minute per 0.05 ml portion of each 10 ml fraction in Bray's solution; the continuous line represents absorbancy per cm at 260 cm. Fractions were eluted with a solution containing a linear gradient of NaCl (0.30 to 0.75 M). Approximately 95% of the applied TCA precipitable cpm were recovered.

species reveals that the relative template activity of synonym codons is affected by the relative abundance of certain Arg-tRNA species, multiplicity of Arg-tRNA species recognizing certain codons, and finally the presence or absence of certain Arg-tRNA species. For example, CGC, a weak template for unfractionated (*E. coli*) Arg-tRNA and strong template for unfractionated (liver) Arg-tRNA, is recognized by only one species of (*E. coli*) Arg-tRNA (14% of total (*E. coli*) tRNA^{ARG} acceptance) and three species of (liver) Arg-tRNA (52% of total (liver) tRNA^{ARG} acceptance).

Ser-tRNA preparations from both *E. coli* B and liver were fractionated into four peaks as shown in fig. 3. Peak 1 (*E. coli*) Ser-tRNA responds to UCU and UCC; peak 2 responds to AGU and AGC; and peaks 3 and 4 respond to UCA, UCG, and less well to UCU. The first fraction (liver) Ser-tRNA responds to AGU and AGC and hence resembles *E. coli* fraction 2. The second smaller fraction responds to UCG. The third (liver) Ser-tRNA

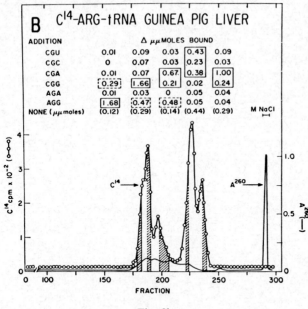

ADDITION		Δ μμMOLES BOUND				
CGU	0.01	0.09	0.03	[0.43]	0.09	
CGC	0	0.07	0.03	0.23	0.03	
CGA	0.01	0.07	[0.67]	[0.38]	[1.00]	
CGG	[0.29]	[1.66]	0.21	0.02	0.24	
AGA	0.01	0.03	0	0.05	0.04	
AGG	[1.68]	[0.47]	[0.48]	0.05	0.04	M NaCl
NONE (μμmoles)	(0.12)	(0.29)	(0.14)	(0.44)	(0.29)	

Fig. 2b

Fig. 2.(b) [14C]-Arg-tRNA from liver (3.4 × 10⁶ cpm precipitated in 5% TCA at 3°C, 264 A²⁶⁰ units) was fractionated by reverse phase column chromatography. Symbols are explained above. Fractions were eluted with a solution containing a linear gradient of NaCl (0.30 to 0.65 M). Approximately 90% of the applied TCA precipitable cpm were recovered.

Δμμmoles represents the amount of [14C]- or [3H]-AA-tRNA bound to ribosomes due to the addition of tri- or polynucleotides. μμmoles represents the amount of [14C]- or [3H]-AA-tRNA bound to ribosomes without the addition of tri- or polynucleotides (enclosed in parentheses).

fraction responds to UCU, UCC, and UCA. A fourth, small, peak of (liver) Ser-tRNA does not respond appreciably to any of the known serine codons; however, small reproducible, responses to poly (U, C) is observed. Since both (liver) and (*E. coli*) Ser-tRNA species recognize alternate codons AGU and AGC their anticodons (ACU) are predicted to be identical. The remaining (liver) and (*E. coli*) Ser-tRNA species differ completely in the alternate codon sets recognized by these molecules and therefore their anticodons. These anticodon mutational events which have occurred during the evolution of the mammalian cell from lower forms altered the sets of synonym codons recognized by Ser-tRNA isoaccepting species without effecting amino-acid–RNA code relationships. Thus strong evolutionary pressure is consistently

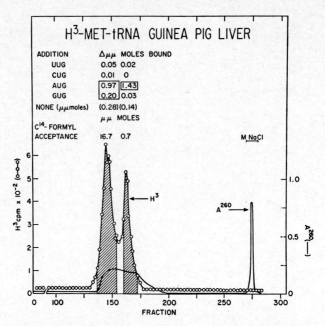

Fig. 3.(a) [14C]-Ser-tRNA from *E. coli* B (3.1 × 10⁶ cpm precipitated in 5% TCA at 3 °C, 302 A²⁶⁰ units) was fractionated by reverse phase column chromatography. The symbols (o--o--o) represent the total cpm per 0.05 ml portion of each 10 ml fraction in Bray's solution and the continuous line represents absorbance/cm at 260 mμ. Fractions were eluted with a solution containing a linear gradient of NaCl (0.30 to 0.75 M). The last Ser-tRNA was eluted by washing the column with a solution containing M NaCl. Approximately 62% of the applied TCA precipitable cpm were recovered.

observed for the universality of the amino acid code with some freedom permitted within the tRNA code.

Two peaks of (liver) Met-tRNA are separated as shown in fig. 4. At 0.01 M Mg⁺⁺, the first (liver) Met-tRNA responds to AUG and GUG. The second peak responds only to AUG. In the presence of purified preparations of *E. coli* transformylase only peak 1 (liver) Met-tRNA accepted formyl groups from $N^{5,10}$ THFA to form N-formyl-Met-tRNA [22]. Neither N-formyl-Met-tRNA nor transformylase activity were detected in liver extracts however. Peak 1 (liver) Met-tRNA bears a third similarity to the (*E. coli*) N-formyl-Met-tRNA species which is involved in initiation of *E. coli* protein synthesis. As shown in part B of fig. 4, the relative affinity of peak 1 and peak 2 (liver) Met-tRNA for initiator sites on *E. coli* ribosomes was studied by determining the rate of methionyl-puromycin [20] from ribosomal bound Met-tRNA. In the presence of AUG, methionyl-puromycin is formed at a

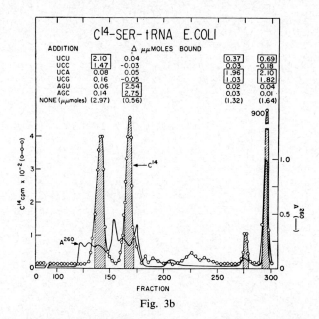

Fig. 3b

Fig. 3.(b)   [³H]-Ser-tRNA from guinea pig liver (5.1 × 10⁶ cpm precipitated in 5 % TCA at 3 °C, 317 A²⁶⁰ units) were fractionated by reverse phase column chromatography. The symbols are described above. Fractions were eluted with a solution containing a linear gradient of NaCl (0.35 to 0.75 M). Approximately 60% of the applied TCA precipitable cpm were recovered.

$\Delta\mu\mu$moles represents the amount of [¹⁴C]- or [³H]-AA-tRNA bound to ribosomes due to the addition of tri- or polynucleotides. $\mu\mu$moles represents the amount of [¹⁴C]- or [³H]-AA-tRNA bound to ribosomes without the addition of tri- or polynucleotides (enclosed in parentheses).

faster rate with peak 1 (liver) Met-tRNA than peak 2 (liver) Met-tRNA. These results suggest that peak 1 (liver) Met-tRNA has a high affinity for initiator sites on *E. coli* ribosomes. Thus several structural features of the (*E. coli*) N-F-Met-tRNA, the tRNA species which initiates protein synthesis in *Escherichia coli* [23] have been retained during the evolution of mammals from lower forms: (1) recognition of mRNA codons which vary in the 1st base position of the codon [24, 25]; (2) structural features which permit recognition by *E. coli* transformylase [26]; (3) affinity for ribosomal sites of initiation [26]. These indirect studies suggest peak 1 (liver) Met-tRNA has several properties of an initiating species of aminoacyl-tRNA and could be involved in the initiation of mammalian protein synthesis.

To explore the possibility that aminoacyl-tRNA differences occur not only at the inter-cellular but also at the subcellular level, mammalian liver was

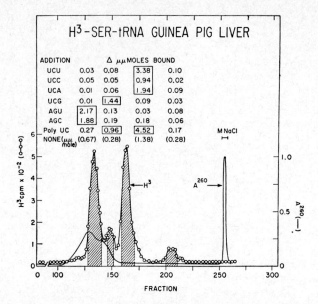

Fig. 4.(a) [³H]-Met-tRNA of liver (3.4 × 10⁶ cpm precipitated in 5% TCA at 3°C and 360 A²⁶⁰ units) was fractionated by reverse phase column chromatography. (0--0--0) represents total [¹⁴C]- or [³H]-counts per minute per 0.05 ml portion of each 10 ml fraction in Bray's solution; the continuous line represents absorbancy/cm at 260 mμ. Fractions were eluted with a solution containing a linear gradient of NaCl (0.35 to 0.70 M). Approximately 82% of the applied TCA precipitable cpm were recovered.

separated by differential centrifugation in sucrose into nuclear, mitochondrial, and cytoplasmic fractions [27]. Transfer RNA was isolated from each fraction, and [¹⁴C]- or [³H]-aminoacyl-tRNA prepared with each by a single aminoacyl-tRNA synthetase which was partially purified from whole liver, and devoid of endogenous tRNA. Following preparation of each nuclear, mitochondrial, and cytoplasmic [¹⁴C]- or [³H]-aminoacyl-tRNA were cochromatographically by RPCC and double label counting techniques. Most aminoacyl-tRNA fractions did not differ from each other in relative abundance or number of isoaccepting species. Certain differences were consistently observed, however.

In the upper portion of fig. 5 is shown the cochromatography of (cytoplasmic) and (nuclear) Lys-tRNA. The radioactive scales were adjusted so that the tritium and carbon-14 values for peak two coincide. Cytoplasm and nuclei contain both isoaccepting species of Lys-tRNA. Nuclei contain a low amount of Lys-tRNA which respond to the codons AAA and AAG. In the lower portion of fig. 5 cochromatography of cytoplasmic and mitochondrial

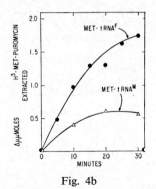

Fig. 4b

Fig. 4.(b) Each 0.050 ml reaction was extracted with ethyl acetate and 0.1 M sodium phosphate, pH 8.0 to determine [^{3}H]-Met-puromycin formation as described by Leder and Bursztyn (1966). $\mu\mu$moles (0--0--0) are corrected for [^{3}H]-Met-puromycin formed in the absence of ApUpG. Each reaction contained: 2.0 A^{260} units of *E. coli* ribosomes; 0.10 A^{260} ApUpG, 0.001 M puromycin, 0.01 M magnesium acetate, 0.05 M potassium acetate, 0.05 M tris acetate pH 7.2, 5.45 $\mu\mu$moles of liver [^{3}H]-Met-tRNAg (fraction 1) or 8.6 $\mu\mu$moles liver [^{3}H]-Met-tRNA$_M$ (fraction 2).

$\Delta\mu\mu$moles represents the amount of [^{14}C]- or [^{3}H]-AA-tRNA bound to ribosomes due to the addition of tri- or polynucleotides. $\mu\mu$moles represents the amount of [^{14}C]- or [^{3}H] AA-tRNA bound to ribosomes without the addition of tri- or polynucleotides (enclosed in parentheses).

Lys-tRNA did not reveal significant differences in the ratio of the two iso-accepting species.

In the upper portion of fig. 6 the cochromatography of (cytoplasmic) and (mitochondrial) Gly-tRNA is shown. It is apparent that the mitochondrial fraction contains a Gly-tRNA species not detected in the cytoplasmic fraction. This unusual Gly-tRNA species contributes 70 per cent of the total (mitochondrial) Gly-tRNA. The relative abundance of the two additional isoaccepting Gly-tRNA peaks corresponds to that of the cytoplasmic fraction. In the lower portion of fig. 6 cochromatography of (cytoplasmic) and (nuclear) Gly-tRNA demonstrates a low level in the nuclear fraction of the Gly-tRNA species abundant in the mitochondria. No detectable level of the first Gly-tRNA peak was observed in the cytoplasmic fraction. Therefore, mitochondria contain 7–8 times the amount of Gly-tRNA$_1$ relative to the isoaccepting species of Gly-tRNA common to all fractions as that found in the nuclear fraction. Since both microscopic and enzymic evidences indicate mitochondrial contamination of the nuclear fraction, it is believed that the low level of Gly-tRNA$_1$ found in the nuclear fraction is mitochondrial contamination. The data suggest guinea pig liver mitochondria contain a

164 C. THOMAS CASKEY

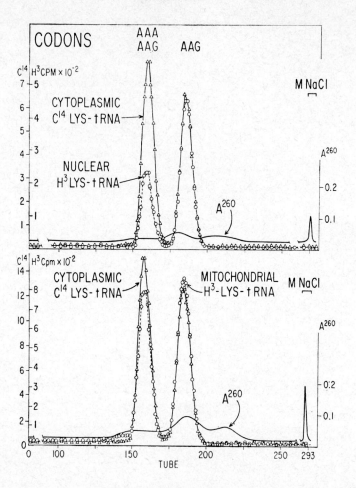

Fig. 5.   (Upper) A mixture of nuclear [³H]-Lys-tRNA (1.1 × 10⁴ cpm precipitated in 5%
TCA at 3°C), cytoplasmic [¹⁴C]-Lys-tRNA (2.8 × 10⁴ cpm precipitated in 5% TCA at
3°C), and 5 mg (*E. coli*) tRNA (125 A²⁶⁰ units) were cochromatographed by RPCC.
(△–△) and (○–○) represent total corrected cytoplasmic [¹⁴C]- and nuclear [³H]-Lys-
tRNA precipitated in 5% TCA at 3°C for each 10 ml fraction; the continuous line represents absorbancy per cm at 260 mμ. Fractions were eluted with a solution containing a
linear gradient of NaCl (0.4 to 0.8 M).
   (Lower) A mixture of mitochondrial [³H]-Lys-tRNA (2.9 × 10⁴ cpm precipitated in 5%
TCA at 3°C); cytoplasmic [¹⁴C]-Lys-tRNA (2.4 × 10⁴ cpm precipitated in 5% TCA at
3°C), and 5 mg (*E. coli*) tRNA (125 A²⁶⁰ units) were cochromatographed by RPCC.
(△–△) and (○–○) represent total corrected cytoplasmic [¹⁴C]- and mitochondrial [³H]-
Lys-tRNA precipitated in 5% TCA at 3°C for each 10 ml fraction; the continuous line
represents absorbancy per cm at 260 mμ. Fractions were eluted with a solution containing
a linear gradient of NaCl (0.4 to 0.8 M).

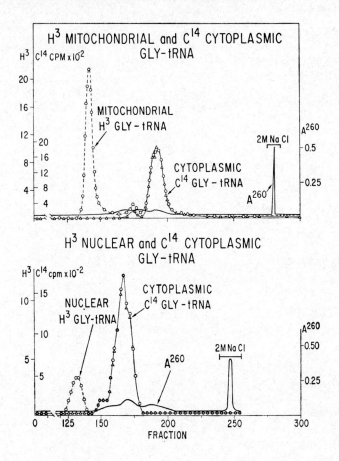

Fig. 6. (Upper) A mixture of mitochondrial [³H]-Gly-tRNA (2.8 × 10⁴ cpm precipitated in 5% TCA at 3 °C), cytoplasmic [¹⁴C]-Gly-tRNA (5.2 × 10⁴ cpm precipitated in 5% TCA at 3 °C), and 5 mg (*E. coli*) tRNA (125 A²⁶⁰ units) were cochromatographed by RPCC. (△ − △) and (○ − ○); represent total corrected cytoplasmic [¹⁴C]- and mitochondrial [³H]-Gly-tRNA precipitated in 5% TCA at 3 °C for each 10 ml fraction; the continuous line represents absorbancy per cm at 260 mμ. Fractions were eluted with a solution containing a linear gradient of NaCl (0.35 to 0.80 M).

(Lower) A mixture of nuclear [³H]-Gly-tRNA (3.0 × 10⁴ cpm precipitated in 5% TCA at 3 °C). Cytoplasmic [¹⁴C]-Gly-tRNA (3.0 × 10⁴ cpm precipitated in 5% TCA at 3 °C), and 5 mg (*E. coli*) tRNA (125 A²⁶⁰ units) were cochromatographed by RPCC. (△ − △) and (○ − ○) represent total corrected cytoplasmic [¹⁴C]- and nuclear [³H]-Gly-tRNA precipitated in 5% TCA at 3 °C for each 10 ml fraction; the continuous line represents absorbancy per cm at 260 mμ. Fractions were eluted with a solution containing a linear gradient of NaCl (0.35 to 0.80 M).

Gly-tRNA species unique to that organelle. Mitochondria from rat liver and *Neurospora crassa* similarly have been found to contain certain species of aminoacyl-tRNA not common to the cytoplasm of these cells [28, 29].

Three fractions of cytoplasmic (liver) Gly-tRNA are identified and shown in fig. 7. Fractions 1 and 2 respond to GGA and poly A: G. Fraction 3

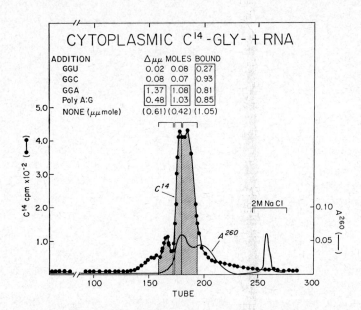

Fig. 7. [14C]-Gly-tRNA cytoplasmic fraction of liver (2.5×10^6 cpm precipitated in 5% TCA at 5°C and 550 A^{260}) was fractionated by reverse phase column chromatography. (●-●-●) represents total [14C] counts per min per 0.05 portion of each 10 ml fraction in Bray's solution; the continuous line represents absorbancy/cm at 260 mμ. Fractions were etuted with a solution containing a linear gradient of NaCl (0.35 to 0.80 M). $\Delta\mu\mu$-moles represents the amount of [14C]-Gly-tRNA bound to ribosomes due to the addition of tri- or polynucleotides. $\mu\mu$moles represents the amount of [14C]-Gly-tRNA bound to ribosomes without the addition of tri- or polynucleotides (enclosed in parentheses).

responds to GGU, GGC, GGA, and poly A: G. The three mitochondrial fractions corresponding to these cytoplasmic Gly-tRNA species recognized identical sets of synonym codons. Mitochondrial Gly-tRNA$_1$ however did not recognize any of the glycine codons or any of the polynucleotides given in table 3. In experiments not shown here this unusual tRNA was shown not to differ from other Gly-tRNA species in molecular weight, linkage of glycine to terminal adenosine residue, recognition by glycine aminoacyl-tRNA

TABLE 3
Mitochondrial Gly-tRNA₁ RNA codon recognition

Oligo- or polynucleotide	$\Delta\mu\mu$moles [³H]-Gly-tRNA Bound to ribosomes
GGU, GGC, GGA	< 0.05
Poly (A, G), (U, G), (C, G)	< 0.01
Poly U, C, A, G	< 0.09
Poly (U, C), (U, A), (C, A)	< 0.01
Poly (U, C, A), (U, C, G), (A, C, G), (A, G, U)	< 0.01
None ($\mu\mu$moles)	(0.20)

Template activity of mRNA codons for guinea pig liver (mitochondrial) Gly-tRNA₁. Each 0.050 ml reaction contained tris acetate 0.05 M pH 7.2; potassium acetate 0.05 M; magnesium acetate 0.02 M; 1.5 A²⁶⁰ units of *E. coli* B ribosomes; 3.21 $\mu\mu$moles [³H]-Gly-tRNA₁ 0.150 A²⁶⁰ units of trinucleotide; and 0.250 A²⁶⁰ units of polynucleotide. All mRNA templates were active for directing the ribosomal binding of appropriate aminoacyl-tRNAs under the conditions employed as determined by parallel studies.

synthetase, and further did not interfere with codon recognition of unfractionated Gly-tRNA. Although the function of this unusual species of tRNA is unresolved at present it should be pointed out that aminoacyl-tRNA can be involved in synthesis of biologic compounds not requiring ribosomes or mRNA. Recently tRNA has been shown to act as an intermediate in synthesis of phospholipids [30, 31], peptidoglycan [32] independent of ribosomes, and activated amino acids occur as intermediates in the synthesis of small peptides [33] independent of ribosomes. All of the known examples occur in bacteria. Since the mitochondria because of its circular DNA [34], 70S ribosomes [35], and other features bear so many similarities to bacteria, this Gly-tRNA species may be acting in synthesis of certain mitochondrial products which occur independently of mRNA.

These comparative studies indicate aminoacyl-tRNA species from different cellular and subcellular sources differ in many ways: (1) relative abundance; (2) numbers of species; (3) sets of alternate synonym codons recognized by aminoacyl-tRNA species resulting from differing tRNA anticodons. These cellular differences in tRNA correlate well with differences in relative template activity of synonym codon originally observed with unfractionated (liver) and (*E. coli*) aminoacyl-tRNA. These and other studies [20, 36] further point out that isoaccepting species of bacterial and mammalian aminoacyl-tRNA can recognize one or more mRNA codons. The alternate codon sets recognized by a single adaptor molecule have been determined experimentally in studies such as these and discussed on a theoretical basis in the Wobble

Hypothesis [37]. It is now possible to relate the patterns of alternate codon recognition to the bases contained in the tRNA anticodon loop since the complete structure of a number of tRNA molecules is now known [38–40]. The number of alternate codon sets recognized by a single molecular species are limited as shown in table 4. Alternate codons which differ in the third

TABLE 4

Alternate codon recognition types

Type	Base position of codon			Anticodon	
	5′	Middle	3′	Adaptor	Base
IA	A=G			F-Met-tRNA	U
IB		A=G		Terminator R_2	Unknown
IC			A=G	Terminator R_1	Unknown
				AA-tRNA	U
II			U=C	AA-tRNA	G
III			U=C=A	AA-tRNA	I
IV			U=G=A	AA-tRNA	U*
V			G	AA-tRNA	C

* Uracil; pseudouracil, 4-thio-uracil.

(3′-position) base position of the triplet mRNA codon are most common (five types). The species of tRNA involved in initiation of E. coli protein synthesis is the sole example of alternate codon recognition of mRNA codons which differ in the first base position (5′-position). Similarly, the terminator adaptor molecule R_2 is the only example of alternate codon recognition of mRNA codons which differ in the middle base position. Types IA, IC, II, III, and V were identified for both mammalian and bacterial aminoacyl-tRNA species, while type IV has only been identified for bacterial aminoacyl-tRNA. The frequency of type III appears high for mammalian aminoacyl-tRNA (5 examples for 10 amino acids studied) and low for E. coli aminoacyl-tRNA (2 examples for 18 amino acids studied). The trypsin sensitive, RNAse insensitive terminator molecules R_1 and R_2 have been described for E. coli [41] and the chemical composition of their anticodons is undetermined.

A summary of a partial tRNA code for E. coli, yeast, and mammals is given in fig. 8. Single adaptor molecules are indicated by joined symbols adjacent the alternate codons which these recognize. Thirty-one species of aminoacyl-tRNA and two terminator molecules have been identified from E. coli corresponding to 14 amino acids and peptide chain termination

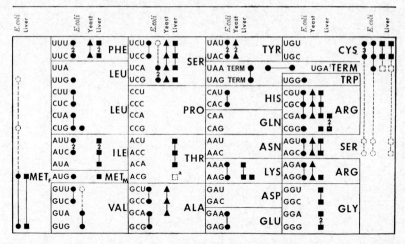

Fig. 8. A summary of the results obtained in this laboratory with purified AA-tRNA fractions is shown in this figure. Additional results have been reported by Soll, Cherayil and Bock [36]. Synonym codon sets were determined by stimulating the binding of purified fractions of AA-tRNA from *E. coli*, yeast, or guinea pig liver fractions to *E. coli* ribosomes with trinucleotide codons. The joined symbols adjacent to the codons represent synonym codons recognized by one purified AA–tRNA fraction from (●, *E. coli*; ▲, yeast; or ■, guinea pig liver). The number between symbols represents the number of redundant peaks of AA–tRNA found which respond similarly to codons. The open symbols represent AA–tRNA responses which may be ambiguous. Results are from Doctor et al. [25]; Kellogg et al. [25]; Nirenberg et al.; Sueoka, Nirenberg and Sueoka, (in press); and Scolnick et al. [41]. Results with Gly-tRNA (liver) and Phe-tRNA (liver) were obtained by C. T. Caskey; Ser-tRNA (Yeast), C. Allende; AGA-AGG-Arg-tRNA (*E. coli*) and Arg-tRNA (Yeast), F. Anderson; His-tRNA (*E. coli*), M. Nirenberg; and Glu-tRNA (*E. coli*), M. Wilcox. ([a]: Uncertain.)

function respectively. An additional minimal number of 9 aminoacyl-tRNA species for the six remaining amino acids can be predicted. Twenty-seven (liver) aminoacyl-tRNA species were identified for 10 amino acids, and an additional minimal number of 13 species predicted for the remaining 10 amino acids. Therefore the (*E. coli*) and (liver) tRNA code have a minimum of 40 adaptor species for 20 amino acids. The relative abundance, number, and anticodons of isoaccepting aminoacyl-tRNA from various cell sources differ however.

These differences in cellular and intracellular organelle content aminoacyl-tRNA between cell types represent changes in codon recognition apparatus acquired after cells had evolved and may enable cells to regulate additional genetic information and/or be a part of differentiation. These differences in cellular content of aminoacyl-tRNA are quite analogous to those which occur when suppressor genes are present, and therefore may be influencing the rate of protein synthesis, synthesis of specific proteins, or effecting the level of translational error. It is clear however that the potential for translational regulation by aminoacyl-tRNA does exist. The definitive regulatory role of such aminoacyl-tRNA difference on protein synthesis must await the outcome of additional study.

These studies support the following conclusions: (1) The amino acid code of widely divergent cell types is *universal*; (2) Synonym codons have differing template activity for aminoacyl-tRNA from bacteria, mammals, and amphibians; (3) Similar patterns of alternate codon recognition are observed with isoaccepting (*E. coli*) and (guinea pig liver) aminoacyl-tRNA while the actual tRNA code of the two cell types differs.

These studies were conducted in collaboration with Dr. R. Marshall, Dr. A. Beaudet, and Dr. M. Nirenberg. Mrs. N. Heaton, Mrs. A. Baur, and Mrs. T. Caryk assisted in the studies. All abbreviations are described in the following references: [13, 20].

References

[1] M. Nirenberg and P. Leder, Science *145* (1964) 1399.
[2] P. Leder and M. Nirenberg, Proc. Natl. Acad. Sci. U.S. *52* (1964) 420.
[3] P. Leder and M. Nirenberg, Proc. Natl. Acad. Sci. U.S. *52* (1964) 1521.
[4] J. Trupin, F. Rottman, R. Brimacombe, P. Leder, M. Bernfield and M. Nirenberg, Proc. Natl. Acad. Sci. U.S. *53* (1965) 807.
[5] M. Bernfield and M. Nirenberg, Science *147* (1965) 479.
[6] D. Söll, E. Othsuka, D. Jones, R. Lohrmann, H. Hayatsu, S. Nishimura and H. G. Khorana, Proc. Natl. Acad. Sci. U.S. *54* (1965) 1368.
[7] M. Nirenberg, P. Leder, M. Bernfield, R. Brimacombe, J. Trupin, F. Rottman and C. O'Neal, Proc. Natl. Acad. Sci. U.S. *53* (1965) 1161.
[8] R. Brimacombe, J. Trupin, M. Nirenberg, P. Leder, M. Bernfield and T. Jaouni, Proc. Natl. Acad. Sci. U.S. *54* (1965) 954.
[9] M. Weigert and A. Garen, Nature *206* (1965) 992.
[10] S. Brenner, A. Stretton and S. Kaplan. Nature *206* (1965) 994.
[11] S. Brenner, L. Barnett, E. R. Katz and F. H. C. Crick, Nature *213* (1967) 449.
[12] C. T. Caskey, R. Tompkins, E. Scolnick and M. Nirenberg, Science *162* (1968) 135.
[13] R. E. Marshall, C. T. Caskey and M. Nirenberg, Science *155* (1967) 820.
[14] E. R. Signer, J. R. Beckwith and S. Brenner, J. Mol. Biol. *14* (1965) 153.
[15] A. Garen, S. Garen and R. C. Wilhelm, J. Mol. Biol. *14* (1967) 167.
[16] J. F. Sambrook, D. P. Fan and S. Brenner, Nature *214* (1967) 452.
[17] J. Carbon, P. Berg and C. Yanofsky, Proc. Natl. Acad. Sci. U.S. *56* (1966) 764.
[18] M. Capecchi and G. Gussin, Science *149* (1965) 417.
[19] C. L. Stebbins, Processes of organic evolution (Prentice-Hall, Englewood Cliffs, N. J., 1966) p. 136.
[20] C. T. Caskey, A. Beaudet and M. Nirenberg, J. Mol. Biol. *37* (1968) 99.
[21] A. D. Kelmers, G. D. Novelli and M. P. Stulberg, J. Biol. Chem. *240* (1965) 3979.
[22] C. T. Caskey, B. Redfield and H. Weissbach, Arch. Biochem. Biophys. *120* (1967) 119.

[23] R. E. Webster, D. L. Englehardt and N. D. Zinder, Proc. Natl. Acad. Sci. U.S. 55 (1966) 155.
[24] B. F. C. Clark and K. A. Marker, J. Mol. Biol. 17 (1966) 394.
[25] D. A. Kellogg, B. P. Doctor, J. E. Loebel and M. W. Nirenberg, Proc. Natl. Acad. Sci. U.S. 55 (1966) 912.
[26] B. F. C. Clark and K. A. Marcker, Nature 211 (1966) 378.
[27] C. T. Caskey, paper in preparation.
[28] W. E. Barnett and D. H. Brown, Proc. Natl. Acad. Sci. U.S. 57 (1967) 452.
[29] C. A. Buck and M. M. K. Nass, Proc. Natl. Acad. Sci. U.S. 60 (1968) 1045.
[30] J. A. Nesbitt and W. J. Lennarz, J. Biol. Chem. 243 (1968) 3088.
[31] R. M. Gould, M. P. Thornton, U. Liepkalns and W. J. Lennarz, J. Biol. Chem. 243 (1968) 3096.
[32] R. M. Bunsted, J. L. Dahl, D. Soll and J. W. Strominger, J. Biol. Chem. 243 (1968) 779.
[33] W. Gevers, H. Kleinkauf and F. Lipmann, Proc. Natl. Acad. Sci. U.S. 60 (1968) 269.
[34] J. H. Sinclair and B. J. Stevens, Proc. Natl. Acad. Sci. U.S. 56 (1966) 508.
[35] K. K. Tewari and S. G. Wildman, Proc. Natl. Acad. Sci. U.S. 59 (1968) 569.
[36] D. Söll, J. D. Cherayil and R. M. Bock, J. Mol. Biol. 29 (1967) 97.
[37] F. H. C. Crick, J. Mol. Biol. 19 (1966) 548.
[38] R. W. Holley, J. Apgar, G. A. Everett, J. T. Madison, M. Marquisse, S. H. Merrill, J. P. Penswick and A. Zamir, Science 147 (1965) 1462.
[39] H. G. Zachau, D. Dutting and H. Feldman, Angew. Chemie Intern. Ed. 5 (1966) 422.
[40] J. T. Madison, G. A. Everett and H. K. Kung, Cold Spring Harbor Symp. Quant. Biol. 31 (1966) 409.
[41] E. Scolnick, R. Tompkins, C. T. Caskey and M. Nirenberg, Proc. Natl. Acad. Sci. U.S. 61 (1968) 768.

EVOLUTION OF PHOSPHAGEN PHOSPHOKINASES

NGUYEN VAN THOAI

Laboratoire de Biochimie Générale et Comparée, Collège de France, Paris, France

Phosphagen phosphokinases are among these proteins which are well fitted to the biochemical evolution analysis. The comparison of some of their features may provide data useful to the knowledge of the natural variations of proteins.

1. The well delimited biological distribution of phosphagens lends itself to phylogenic investigations. It was well known that creatine kinase characterizes vertebrates and arginine kinase, invertebrates. Owing to results gained in our laboratory [1] it is now known that, in Annelida and related phyla, besides these enzymes, other phosphokinases of the same type, are responsible for the formation of five less well known phosphagens: phosphoglycocyamine, phosphotaurocyamine, phosphohypotaurocyamine, phospholombricine and phosphoopheline (fig. 1).

The chemical phylogeny, based on the structure of phosphagens and their respective enzymes may be studied through various groups of worms [2].

According to table 1 it is noticeable that, despite some overlappings, the distribution of phosphagens in worms presents a certain evolutionary continuity. Consequently the trail of the biochemical evolution of phosphagens may be connected with the phylogeny of worms.

2. The similarity of the primary structure of proteins is a good reference to their homology and the more or less important variations shown in the sequential structure would give evidence of the rate of the biochemical evolution. Since phosphagen kinases, the molecular weight of which range generally from 40,000 to about 80,000 [3], are small sized, the determination of their primary structure is quite feasible, specially with the new Edman degradation method.

3. Lastly another feature of phosphagen kinases is fitted to the biochemical evolution study. These enzymes act on two substrates, the one of which, ATP, is always the same and the other presents various, sometimes important, changes. Three of these enzymes, the kinases of arginine, creatine and glycocyamine, are strictly specific, and four of them (the kinases of taurocyamine, hypotaurocyamine, lombricine and opheline) however inactive on the substrates of the first enzymes, are shown to be more or less interspecific [2].

These facts allow to think that the ATP binding site does not change during phosphagen kinases evolution while the guanidine substrates binding sites are more or less modified according to the extent of the guanidine derivatives change. The modifications may consist, by example, in deleting the hydrophobic area primitively required to bind arginine or in adding polar

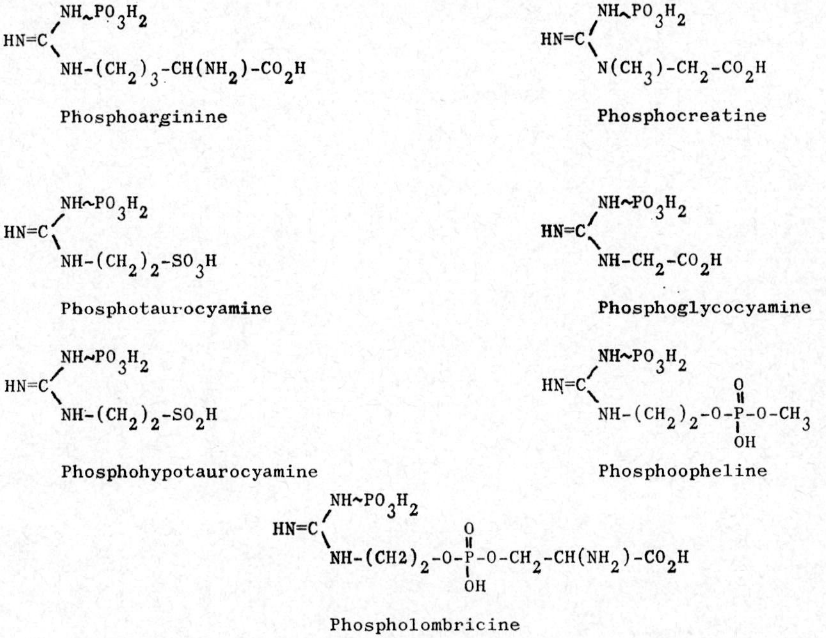

Phosphoarginine

Phosphocreatine

Phosphotaurocyamine

Phosphoglycocyamine

Phosphohypotaurocyamine

Phosphoopheline

Phospholombricine

Fig. 1. Phosphagens.

TABLE 1

Biological distribution of guanidine compounds and phosphagens among Annelida, Echiuroidea, Nemertea, Phoronidea and Sipunculidea

Family	Species	Phosphagens	
POLYCHAETA			
Nereidiformia			
Nereidae	*Nereis brandti*	PG	(PC)
	N. diversicolor	PG	(PC)
	N. fucata	PG	(PC)
	N. limnicola	PG	(PC)
	N. succinea		PC
	N. vexillosa	(PG)	PC
	Perinereis cultrifera	PG	
	Platynereis agassizii	PG	(PC)
	P. dumerilii	PG	

Table 1 (Continued)

Family	Species		Phosphagens	
Nephthydae	*Nephthys californiensis*		PG	
	N. coeca		PG	
	N. coecoides		PG	
	N. hombergii		PG	
Phyllodocidae	*Eulalia viridis*		PG	
	Phyllodoce maculata			PC
Aphroditidae	*Arcatnoë pulchra*		PG	
	Halosydna brevisetosa		PG	(PC)
	H. gelatinosa		PG	
	H. johnsoni		PG	
	Hesperonoë adventor		PG	
	Sigalion mathildae		PG	
	Sthenelais boa		PG	(PC)
	S. fusca			PC
	S. leidyi			PC
	Aphrodite aculeata			PC
	Harmothoë imbricata			PC
	Hermione hystrix			PC
Amphinomidae	*Pareurythoë californica*			PC
Glyceridae	*Glycinde armigera*			PC
	Glycera americana			PC
	G. convoluta			PC
	G. dibranchiata			PC
	G. gigantea			PC
	G. robusta			PC
	G. tesselata			PC
	Hemipodus borealis			PC
Eunicidae	*Arabella iricolor*			PC
	A. semimaculata			PC
	Diopatra cuprea			PC
	D. napoleitana			PC
	D. splendidissima			PC
	Eunice harassi			PC
	E. sp.			PC
	Halla parthenopeia			PC
	Lumbriconereis erecta	(PA)		PC
	L. impatiens			PC
	L. zonata		(PG)	PC
	Marphysa sanguinea			PC
Ariciidae	*Aricia foetida*			PC
	A. latreillii			PC
	Haploscoloplos elongata			PC
	Scoloplos acmeceps			PC
	S. armiger			PC
Capitelliformia				
Capitellidae	*Dasybranchus caducus*			PC
	D. lumbricoides			PL
	Notomastus tenuis	(PA)		PL

Table 1 (Continued)

Family	Species		Phosphagens		
Oweniidae	*Owenia fusiformis*	PA			
Terebelliformia					
Cirratulidae	*Audouinia tentaculata*		PC		
	Cirriformia luxuriosa		PC		
	C. spirabrancha		PC		
Terebellidae	*Amphitrite edwardsii*	PA		(PT)	
	A. gracilis	PA		(PT)	
	Lanicea conchilega	PA			
	Pista elongata			PT	
	Polymnia nebulosa	(PA)		PT	
	Streblosoma bairdi	PA			
	Terebella californica	PA			
	T. lapidaria	PA	(PC)	(PT)	
	Terebellides stroemi		PC		
	Thelepus cincinnatus	PA			
	T. crispus	PA		(PT)	
	T. setosus	PA			
Sabelliformia					
Serpullidae	*Apomatus similis*	PA		PT	
	Hydroides sp.			PT	
	Mercierella enigmatica	(PA)	(PC)	PT	
	Pomatoceros triqueter		PC		
	Protula intestinum		PC		
	Serpula vermicularis			PT	
	Spirorbis borealis		PC		
Sabellidae	*Bispira sp.*	PA			
	Eudistylia polymorpha		PC		
	Myxicola infundibulum			PT	
	Sabella pavonina	PA			
	Spirographis spallanzanii	PA			
Sabellariformia					
Sabellaridae	*Sabellaria alveolata*			PT	
Spioniformia					
Spionidae	*Nerine cirratulus*			PT	
Scoleciformia					
Arenicolidae	*Arenicola assimilis*			PT	
	A. ecaudata			(PT)	PH
	A. marina			PT	(PH)
Maldanidae	*Axiothella rubrocincta*	(PA)		PT	
	Clymene lumbricoides			PT	
	Leiochone clipeata			PT	
Opheliidae	*Ophelia bicornis*				PL
	O. limacina	(PA)			PL
	O. neglecta				PO
	O. radiata				PL
	Thoracophelia mucronata				PL
	Travisia forbesii	PA	PC		
	T. gigas	(PA)			PL

Table 1 (Continued)

Family	Species	Phosphagens	
OLIGOCHAETA			
Lumbricidae	*Allolobophora caliginosa*		PL
	A. chlorotica		PL
	A. dubiosa		PL
	A. leoni		<u>PL</u>
	A. rosea		<u>PL</u>
	Dendrobaena platyura var. *depressa*		<u>PL</u>
	Eisenia balatonica		<u>PL</u>
	Lumbricus rubellus		<u>PL</u>
	L. terrestris		PL
	Octolasium cyaneum		PL
	O. lacteum		<u>PL</u>
	O. transpadanum		<u>PL</u>
Megascolidae	*Eutyphoeus waltoni*		<u>PL</u>
	Megascolides cameroni		PL
	Pheretima postuma		<u>PL</u>
HIRUDINEA			
	Erpobdella octoculata		
	Glossiphonia complanata		
	Haemopis sanguisuga		
	Hirudo medicinalis	Undetectable phosphagen	
PHORONIDEA			
	Phoronopsis viridis	<u>PC</u>	
ECHIUROIDEA			
	Bonellia viridis	Undetectable phosphagen	
	Urechis caupo		PL
NEMERTEA			
	Cerebratus occidentalis	<u>PA</u>	
	Lineus pictifrons	<u>PA</u>	
SIPUNCULIDEA			
	Sipunculus nudus	PA	
	Dendrostomum dyscritum	(PT)	PH
	Phascolion strombi	<u>(PT)</u>	<u>PH</u>
	Phascolosoma agassizii	(PT)	PH
	P. elongatum	(PT)	PH
	P. vulgare	(PT)	PH
	Siphonosoma ingens	(PT)	PH

From Thoai and Robin [1].

( ): minor component. Underlined: hypothetic phosphagen, presumed from the content of major guanidine base. PA = phosphoarginine; PC = phosphocreatine; PG = phosphoglycocyamine; PH = phosphohypotaurocyamine; PL = phospholombricine; PO = phosphoopheline; PT = phosphotaurocyamine.

groups used to bind $-CO_2H$, $-SO_3H$ or PO_3H groups. If consequently to various changes of guanidine substrates successive modifications of the enzymes reach a great extent, they may induce further changes in other areas of the proteins so as to adjust the required structure of the catalytic site. It is now questionable if a certain extent of modifications should entail a change of the reaction mechanism.

In the case of phosphagen kinases, we are interested in these two aspects of the evolution: what area of the proteins which are preserved consequently to the fact that one of two substrates, ATP or ADP, keep unchanged and do the phosphagen binding site change markedly? Evidence is given that modifications brought to the last binding site, even not complete, give rise to a change of the reaction mechanism of these homologous enzymes.

Homology of phosphagen kinases

Similar amino acid composition. The first indication of the homology of phosphagen kinases is found in their amino acid composition [4–6]. As is shown in table 2, the percentage of various groups of amino acids (diacids, diamino acids, aromatic acids etc...) is remarkably constant for seven enzymes, otherwise quite different in the respects of the size, the conformation, the specificity etc...

For different categories of amino acids, the deviation from the average figure does not exceed 2 g of residue per cent. Such fastness in the amino acid composition must evolve from an ancestral gene, common to all phosphagen kinases.

Similar peptide maps. The essential SH group of five phosphagen kinases has been labelled with [14]C—N-ethylmaleimide and the peptide maps have been obtained from the tryptic hydrolysates. Whatever their molecular weight may be the sum of peptides is about 50 and is equivalent to the same amount of lysine + arginine residues found in the arginine kinase from lobster (MW 40,000) and equivalent to the half of this for other phosphokinases (MW 80,000) [7, 8]. It is remarkable that a lot of these peptides are samely localized on the finger-prints, specially when the electrochromatogram is sprayed with specific reagents for various amino acids (unpublished results). As is shown in fig. 2 the most radioactive peptides have the same position on the chromatogram.

Similar amino acid sequences around the essential SH groups. The isolation and the determination of the amino acid sequences around the essential SH groups of various phosphagen kinases are being carried out by Der Terrossian

NGUYEN VAN THOAI

TABLE 2

Composition of phosphagen phosphokinases. Percentage of amino acid residues grouped by categories*

Amino acid residues	Average	APK C. pagurus	APK Homarus v.	LPK Lumbricus t.	TPK Arenicola m.	CPK rabbit	APK Siunculus n.	GPK Nephthys coeca
Asp + Glu	21.90	23.75	23.89	19.06	20.11	23.67	20.85	22.00
Lys + Arg + His	19.95	19.81	18.57	19.17	21.68	22.56	20.54	17.32
Tyr + Phe + Trp	11.57	11.81	11.82	11.38	10.50	10.83	12.45	13.57
Met + Cys-SH	4.47	4.23	3.91	2.60	4.93	4.09	5.32	6.23
Ser + Thr	8.22	7.22	9.00	7.68	8.10	8.68	8.83	8.07
Pro	3.60	3.19	2.52	3.93	3.93	4.38	4.04	3.27
Gly + Ala + Val + Ile + Leu	27.80	27.43	27.12	26.93	27.14	26.41	29.35	27.82
−CONH$_2$	0.85	0.85	0.73	0.86	0.86	1.08	0.80	0.75

* g amino acid residues/100 g protein.

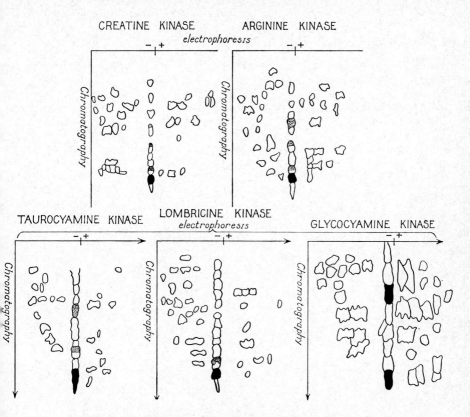

Fig. 2.   Peptide maps of five phosphagen phosphokinases.

and her colleagues. From the first results obtained and as is shown in fig. 3 the amino acid sequence around the essential SH group of lobster arginine kinase is very similar to this of the peptide from rabbit creatine kinase studied by Thomson et al [9] and by Mahowald [10]. It is noticeable that the carboxyl terminal sequence of the two peptides differs only by the deletion of a glycine residue in arginine kinase and of a threonine residue in creatine kinase. The only other difference is the change of a valine residue for a leucine residue what is equivalent to one base mutation [11]. In spite of the strict specificity of arginine and creatine kinases and in spite of the phylogenic remoteness between the rabbit and the lobster, the structure of an essential site is remarkably preserved.

Similar essential amino acid residues. It has been found previously that creatine kinase contains two SH groups [12, 13]. By using 5,5' dithiobis–2-

Arginine kinase (From Der Terrossian et al)

Gl( )-Thr-Cys-Pro-Thr-Ser-Asn-Leu-Gly-Thr-Val-Arg ....
            *
            S-Succ.

Creatine kinase (From Thomson et al.)

Ala-Gly-Pro-His-Phe-Met-(Asp-His-Glu)-(Gly-Leu)-Tyr- Val-Leu-Cys- Pro-Ser-Asn-Leu-Gly-Thr-Gly-Leu-Arg ----
                                                              *
                                                              S-CM

Creatine kinase (From Mahowald)

Val-Leu-Cys-Pro-Ser-Asn-Leu
        *
        S-DNP

Fig. 3. Amino acid sequences around the essential cysteine of arginine- and creatine phosphokinases.

nitrobenzoic acid to titrate stoichiometrically the essential SH groups, we have found that dimeric and large-sized enzymes, the kinases of taurocyamine and glycocyamine, contain, like the creatine kinase, two essential SH groups while the monomeric and small-sized lobster arginine kinase possesses only one essential SH group [7, 8]. Lombricine kinase is an exception: in spite of its molecular weight and of its dimeric structure this enzyme contains only one essential SH group and presents many other arginine kinase features.

Furtherly it has been found by Pradel and Kassab [14] that the kinases of arginine and of creatine are completely inhibited by carbethoxylating their histidine residues with diethylpyrocarbonate. Evidence is given that one histidine residue is blocked in the case of arginine kinase and two in that of creatine kinase.

Evidence that one essential amino acid residue is present by 40,000 g protein unit is given similarly for lysine residue. According to the same workers [15], arginine and creatine kinases are inhibited by dansylating the ε-amino group of lysine. The inhibition is complete when the first enzyme reacts with one mole of 1-dimethylaminonaphtalene-sulfo derivative (dansyl) and the second enzyme, with two moles of the reagent.

Thus for the three essential groups, cysteine, histidine, and lysine, one amino acid residue is present in arginine and creatine kinases by 40,000 g subunit. It is questionable if the passage from the monomeric enzyme to the dimeric one proceeds through exact replication of the first kinase and how this process can secure the specificity of the two enzymes.

Similar changes of the binding site microenvironment. How the essential structures are well preserved can be seen by comparing the variations of the microenvironment induced by the binding of substrates.

Through the difference spectrophotometry method [16] evidence is given that ATP or ADP binding to arginine-, creatine-, lombricine- and taurocyamine kinases gives rises to an identical spectrum. With ATP, at pH 7.15 or 8.5, the spectrum is characterized by a minimum at 254 mμ. In the same conditions, the spectrum produced by ADP is marked by two peaks, respectively at 295 and 278 mμ and by a minimum at 254 mμ. The last spectrum resembles to that produced by ADP-Mg at pH 6.7 versus pH 1.6. and has been ascribed to the protonation of the adenine ring [17]. Furtherly the spectrum produced by ADP binding to phosphagen kinases discloses a specific contribution from the enzymic protein visualized by the peak at 294 mμ, which should be ascribed to a tryptophane residue (figs. 4 and 5). It is remarkable indeed that ATP or ADP binding produces similar structural changes with so specific enzymes belonging to so different phylogenic bran-

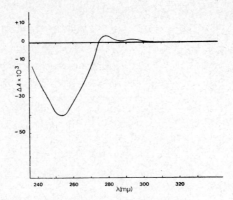

Fig. 4. Difference spectrum produced by arginine- and creatine phosphokinases interacting with ATP (or ATP-Mg).

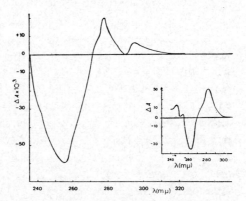

Fig. 5. Difference spectrum produced by arginine- and creatine phosphokinases interacting with ADP (or ADP-Mg).

ches. Moreover it would be stressed that the spectra produced by nucleotidic
substrates with phosphagen kinases are similar to those studied during the
interaction of ADP, ADP-ribose and NAD with alcohol dehydrogenase [18]
or with glutamate dehydrogenase [19]. Nevertheless their greatly different
molecular forms, biochemical evolution has efficiently laid on these enzymes
a convergence of structure at the locus essential to the binding of the same
types of substrates. Despite their complexity and their diversity, biological
macromolecules seem to present the same minimal structures, fitted to the
same catalytic function. These data are meaningful and it is hopeful that the
knowledge of these minimal structures will help to comprehend boht the
mechanism of enzymatic reactions and of the process of biological evolution.

The guanidine substrates binding site has been studied also by the difference spectrophotometry technique. The spectra produced by arginine or phosphoarginine interaction with arginine kinase are characterized by three minima, respectively at 287, 279 and 239 mμ [16] (fig. 6). The first two minima result from a blue shift of the tyrosine residue absorption band and the last

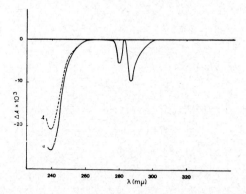

Fig. 6.   Difference spectra produced by arginine phosphokinase interacting with (a) L-arginine and (b) L-arginine plus L-arginine phosphate.

one from the disclosure of a cysteine residue. As a matter of fact, the minimum at 239 mμ is a direct ratio of the amount of N-ethylmaleimide bound to a SH group (unpublished results). This SH group is connected with the binding of arginine by its terminal site $-CH(NH_2)CO_2H$. Indeed the binding of arginine kinase with iodoacetic acid, isoleucine, valine or citrulline produces the same spectrum as with L-arginine (fig. 7). D-arginine, arginic acid, δ-guanidinovaleric acid, γ-guanidinobutane do not give rise to any characteristic spectrum [16].

Creatine kinase and lombricine kinase interaction with their respective substrates, produces no characteristic difference spectrum. But when reacting with iodoacetic acid, the two enzymes give rise to the same spectra observed with arginine kinase. The main feature is that in the first cases the blue shift of the tyrosine residue absorption band is less marked, specially with creatine kinase (unpublished results).

It is obvious that the microenvironment around the essential SH group is preserved in lombricine and creatine kinases, nevertheless this group is not used for the same purpose as in arginine kinase, the binding of the amino acid terminal site $-CH(NH_2)CO_2H$.

From the whole of facts it may be concluded that phosphagen kinases are

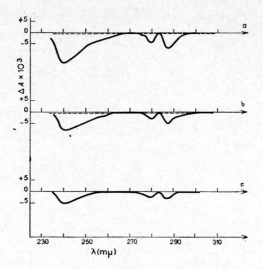

Fig. 7. Difference spectra produced by arginine phosphokinase interacting with (a) isoleucine, (b) valine and (c) citrulline. —— L-aminoacids; ---- D-aminoacids.

homologous proteins. The nature and the number of essential amino acid residues, the sequential structure of the peptide around the essential SH group, the microenvironment surrounding this cysteine and the ATP or ADP binding site as is disclosed by difference spectrophotometry spectra, are well preserved in all studied enzymes.

It may be thought what has been preserved in the course of evolution are the minimal structures of the protein, essential to specific reaction. By fitting their conformation to chemical changes of available substrates, enzymes must have modified some area to a certain extent which has consequently led to a change of the reaction mechanisms.

The divergence of the structure and the change of the reaction mechanism

Besides the similarities which have just been considered, phosphagen kinases present many remarkable differences in the respects of molecular size and conformation.

Different molecular size and conformation. As is shown in table 3, the phosphagen kinases molecular weights range from about 40,000 to 200,000. It is noticeable that arginine kinase is sized from 40,000 to 80,000 and to 160,000, the three molecular weights corresponding to diverse extents of polymerisation. The widest distributed enzyme is the small-sized one [3, 6, 24–26]. The

TABLE 3

Molecular weights of phosphagen phosphokinases

Enzymes	Molecular weights	References
Arginine kinases		
Cancer pagurus	39,000	[6]
Homarus vulgaris	43,200	[3]
Sipunculus nudus	86,000	[20, 21]
Sabella pavonina	160,000	[22]
Creatine kinases		
Rabbit	82,000	[23]
Echinus esculentus, muscle	150,000–185,000	[24]
Echinus esculentus, spermatozoa	> 200,000	[24]
Glycocyamine kinase		
Nephthys coeca	81,000	[3, 8]
Taurocyamine kinase		
Arenicola marina	80,000	[3]
Hypotaurocyamine kinase		
Phascolosoma vulgare	83,000	[3]
Lombricine kinase		
Lumbricus terrestris	74,000	[3]

other two arginine kinases, however obtained in a homogenous state in our laboratory, are less well known [20–22]. Creatine kinase seems to possess samely three different molecular weights [23] but the only one being studied is the small-sized enzyme (MW 82,000). The other four phosphagen kinases have approximatively the same molecular weight as the rabbit creatine kinase. Heavier or smaller size forms are not known. Thus, except for its heavier forms which are not common, arginine kinase ascribed as the most primitive enzyme, is monomeric and the other enzymes which are dimeric would be said to be diverted from the first one.

The molecular conformation of phosphagen kinases varies in the same manner. From table 4 it can be seen that the monomeric enzymes (arginine kinases from crab and from lobster) contain the higher percentage of α-helix structure, approximatively 40%, and the dimeric ones the smaller percentage, ranging from 25 to 30% [27].

Different role of the essential amino acid residues. It has been said previously that phosphagen kinases contain one essential SH group by 40,000 g protein subunit and that the amino acid sequence around the SH group is nearly the

TABLE 4

Content of α-helix structure ($\alpha\%$) in phosphagen phosphokinases evaluated according to optical rotatory dispersion parameters

Enzymes	$\alpha\%_{\lambda c}$	$\alpha\%_{ob_0}$	$\alpha\%$ A_{195}–A_{225}	$\alpha\%_{ob_0}$ *
Arginine kinase MW 39,000	39	37	42	38
Arginine kinase MW 43,200	39	39	45	42
Lombricine kinase	30	28	33	28
Taurocyamine kinase	35	34	39	34
Creatine kinase	31	24		
Glycocyamine kinase	27	27		

* Parameter evaluated according to Schechter–Blout constants and to the relation $b_0 = 0.142\,A_{193} - 0.142\,A_{225}$.

same in arginine and creatine kinases. Further the same difference spectra are produced when iodoacetic acid reacts with the two enzymes and with lombricine kinase.

These data lead to assign to the essential SH group the same role for different phosphagen kinases. Obviously such is not the case.

As a matter of fact, the inhibition of arginine and lombricine kinases by 5,5' dithiobis-2-nitrobenzoic acid is antagonized efficiently by their respective guanidine substrates while creatine and glycocyamine kinases are sheltered from the same reagent by ATP-Mg and ADP-Mg. Taurocyamine kinase is an intermediate case: the enzyme is protected by both nucleotidic and guanidine substrates [17].

From these data it may be thought either the SH group does not play the same role in the three types of enzymes or it is sterically closer in one case to the guadinine binding site and in the other case to the nucleotide binding site.

The same type of facts has been found in the case of the essential histidine residue. The inhibition of arginine kinase by diethylpyrocarbonate is antagonized partially by arginine or by ATP-Mg and much better by the association of the two substrates. On the contrary, creatine kinase cannot be sheltered by its substrates, acting separately or in concert [14]. From other unpublished results obtained by difference spectrophotometry studies and too long to be analyzed presently, evidence is given that histidine is not only used to bind arginine or ATP-Mg but is directly connected with the enzymatic process. Finally it may be stated that histidine residue does not play exactly the same role for arginine kinase and for creatine kinase.

Different reaction mechanisms. The changes in the localization of the same essential amino acid residues, eventually associated with more important modifications in their closeness and in their hydrophobic environment may induce a change in the reaction mechanism.

Kinetic studies carried out on arginine kinase give evidence [unpublished results] that, as with the enzyme from *Jasus verreauxi* [21], their reaction mechanism is a "ping-pong" type mechanism mixed with a sequential type one [29]. On the contrary transphosphorylations catalysed by the kinases of creatine, lombricine and taurocyamine are cooperative types (unpublished results). In the first case, the terminal phosphoryl group should be first unbound by a nucleophilic attack from a histidine residue and finally transferred to the guanidine group of arginine. In the case of other phosphagen kinases, it may be supposed that the transphosphorylation proceeds directly from ATP to the guanidine group of the substrate, consequently, by example, to a slight shift in the localization of the histidine residue. This group still acts for the enzymatic catalysis, but through another way, let us say, to determine the transition state of the β-phosphoryl group of ATP.

Such a change of mechanism may result from very slight modifications of the catalytic site. Small removal or bringing together of essential amino acid residues, in relation to the localization of the substrates, may explain the change from a two step transphosphorylation to a concerted reaction.

Conclusions

The study of phosphagen kinases gives evidence that, during the course of evolution, minimal essential structures are preserved such as the nature and the number of essential amino acid residues, the primary structure of amino acid sequences around certain of these residues, the microenvironment of the substrate binding sites. But in other respects, slight modifications of the localization of the same amino acid residues would suffice to change the transphosphorylation mechanism. Biochemical evolution has led to a divergence of the molecular sizes and conformations, which in its turn, entails a change in the reaction mechanism but the process has maintained minimal structures essential to the type of enzymatic reaction.

References

[1] N. v. Thoai and Y. Robin, in Chemical zoology, M. Florkin and B. T. Sheer, eds., Vol IV (Academic Press, New York 1968) p. 163.
[2] N. v. Thoai in Homologous enzymes and biochemical evolution, N. v. Thoai and J. Roche, eds. (Gordon and Breach, New York 1968) p. 222.
[3] N. v. Thoai, R. Kassab and L.-A. Pradel, Biochim. Biophys. Acta *110* (1965) 532.
[4] E. A. Noltman, T. A. Mahowald and S. A. Kuby, J. Biol. Chem. *237* (1962) 191.

[5] E. Der Terrossian, R. Kassab, L.-A. Pradel and N. v. Thoai, Biochim. Biophys. Acta *122* (1966) 462.

[6] N. v. Thoai, E. Der Terrossian, L.-A. Pradel, R. Kassab, Y. Robin, M. F. Landon, G. Lacombe and N. v. Thiem, Bull. Soc. Chim. Biol. *50* (1968) 63.

[7] R. Kassab, L.-A. Pradel, E. Der Terrossian and N. v. Thoai, Biochim. Biophys. Acta *132* (1967) 347.

[8] L.-A. Pradel, R. Kassab, C. Conlay and N. v. Thoai, Biochim Biophys. Acta *154* (1968) 305.

[9] A. R. Thomson, J. W. Eveleigh and B. J. Miles, Nature *203* (1964) 267.

[10] Th. A. Mahowald, Biochemistry *4* (1965) 732.

[11] E. Der Terrosian, L.-A. Pradel, R. Kassab and N. v. Thoai, Eur. J. Biochem. *11* (1969) 482.

[12] R. E. Benesch, A. H. Lardy and R. Benesch, J. Biol. Chem. *216* (1955) 633.

[13] D. C. Watts, B. R. Rabin and E. M. Crook, Biochem. J. *82* (1962) 412.

[14] L. A. Pradel and R. Kassab, Biochim. Biophys. Acta *167* (1968) 317.

[15] R. Kassab, Cl. Roustan and L.-A. Pradel, Biochim. Biophys. Acta *167* (1968) 308.

[16] Cl. Roustan, R. Kassab, L.-A. Pradel and N. v. Thoai, Biochim. Biophys. Acta *167* (1968) 326.

[17] R. F. Fischer, A. C. Haine, A. T. Mathias and B. R. Rabin, Biochim. Biophys. Acta *139* (1967) 169.

[18] H. Theorell and T. Yonetani, Arch. Biochem. Biophys. *106* (1964) 252.

[19] D. Pantaloni and M. Iwatsubo, Biochim. Biophys. Acta *132* (1967) 217.

[20] N. v. Thoai, N. v. Thiem, G. Lacombe and J. Roche, Biochim. Biophys. Acta *122* (1966) 547.

[21] G. Lacombe, N. v. Thiem and N. v. Thoai, European J. Biochem. *9* (1969) 237.

[22] Y. Robin, C. Klotz, and N. v. Thoai, Biochim. Biophys. Acta *171* (1969) 357.

[23] R. H. Yue, R. H. Palmieri, O. E. Olson and S. A. Kuby, Biochemistry *6* (1967) 3204.

[24] B. Moreland, D. C. Watts and R. Virden, Nature *214* (1967) 468.

[25] P. Elodi and E. Szorényi, Acta Physiol. Acad. Sci. Hung. *9* (1956) 367.

[26] B. Moreland and D. C. Watts, Nature *215* (1967) 1092.

[27] C. Oriol-Audit, M. F. Landon, Y. Robin and N. v. Thoai, Biochim. Biophys. Acta, *188* (1969) 132.

[28] M. L. Uhr, F. Marcus and J. F. Morrison, J. Biol. Chem. *241* (1966) 5428.

[29] W. W. Cleland, Ann. Rev. Biochem. *36* (1967) 77.

FUNCTIONAL ORGANISATION AND MOLECULAR EVOLUTION

D. C. WATTS

Biochemistry Department, Guy's Hospital Medical School, London S.E.1., England

Although catalytic proteins may have formed part of the first organised biological systems it required the emergence of a hereditary genome with its dependent phenotype before life and evolution, as we understand it, could begin [1]. Of the first enzymes, some, by nature of their essential function, would be capable of only limited evolutionary change while others, less essential, might slowly acquire new catalytic functions. However, such progressive modification of existing enzymes provides no possibility for increasing the total functional performance of an organism in terms of additional catalytic ability. This can only come from additional enzymes and hence requires new genes [2].

A polypeptide containing a random arrangement of amino acids would, even under favourable conditions, only slowly acquire a stable conformation and a useful catalytic ability. It follows that the most suitable genetic material for coding a new enzyme is likely to be that which already codes for an enzyme and, preferably, one with a function related to the new catalytic property required. As is now well recognised, such extra genetic material can be generated by duplication of all or part of the genome. Among eukaryotes the *primary* duplication of single genes or small groups of genes is probably only possible by gross chromosomal rearrangement since the alternative process of crossing over during meiosis only follows the pairing of highly homologous regions of the two chromosomes [3]. The importance of homology in pairing at meiosis is well illustrated in the hexaploid wheat, *Triticum aestavum* [4]). Here, although the triplicate genomes of this species contain genes in common, at meiosis the same chromosomes always paired together and the regular formation of bivalents showed that mis-pairing of possibly similar chromosomes did not occur. Minimization of non-homologous pairing is undoubtedly an essential feature of meiosis. However, where chromosomal rearrangement results in duplicate genes being adjacent on the same chromosome arm, there is immediate possibility for further, *secondary*, duplication by homologous but unequal pairing. At the gross character level this is illustrated by the Bar Eye gene in *Drosophila* and at the protein level by the human Lepore haemoglobins [3]. The extent of homology required for such pairing to occur is not known but an indication is given by the deletion mutant, haemoglobin M. Freiberg, which probably arose by crossing over following unequal pairing [5]. Inspection of the amino acid sequence around the deleted valine residue (fig. 1) suggests that identical pairing could only occur between two glycine residues. In contrast, comparison of the codon sequences reveals a 63% identity between the first two nucleotides of

Amino acid sequence
$$(23)$$
Val.Asp.Glu.Val.Gly.Gly.Glu.Ala
$$\times$$
Val.Asp.Glu.Val.Gly.Gly.Glu.Ala
$$(23)$$
$$\downarrow$$
Val.Asp.Glu.Gly.Gly.Glu.Ala   (HbM Freiberg)
$$+$$
Val.Asp.Glu.Val.Val.Gly.Gly.Glu.Ala   (Not known)

Codon alignment
$$-GUX.GAX.GAX.GUX.GGX.GGX.GAX.GCX-$$
$$-GUX.GAX.GAX.GUX.GGX.GGX.GAX.GCX-$$

Fig. 1. Evaluation of the codon alignment as a basis for pairing before unequal crossing over in the deletion mutant, haemoglobin Freiberg. The numbers in parenthesis indicate the position in the polypeptide of the valine which is lost from one gene and gained by the other gene.

each codon (the third nucleotide cannot be known with certainty) over a span of seven amino acids. It would seem that this provides a sufficiently high degree of correspondence for the two genes to pair three nucleotides out of register, at least along part of their length. Because of the nature of the genetic code, this degree of homology is close to that which would occur between two identical genes with different codons for the amino acid at each position in the polypeptide. (In practice random mutation at each codon would result in the degree of homology being nearer 75% because some mutations would produce identical codons at particular positions.) Although there is not yet enough evidence from naturally occurring mutants to test this possibility it is most probable that selection would act against any significant level of degeneracy between homologous genes. However, degeneracy of the code may have two selective advantages in aiding the further evolution of newly duplicated genes, an important feature of which must be that the rapid accumulation of random mutations should not destroy the potentially useful new protein before it can be selectively stabilized by incorporation into the inevitably more slowly changing metabolic environment. In addition to the arrangement of the code being such that single step mutations tend to result in conservative amino acid substitutions, on average one in four mutations would result in an unchanged amino acid with the added benefit that each mutation gathered in the genome will decrease the possibility of unequal pairing between structurally homologous but locationally different genes. The rare combination of properties that allows homologous unequal crossing over to occur occasionally in the Lepore haemoglobins is much less likely to occur for enzymes with their greater diversity of function and, in most cases, the absence of a stabilizing prosthetic group.

The occurrence of duplicate genes does not necessarily mean that they will ultimately be used to code for new enzymes. This requires the matching of the newly acquired enzyme function with some selectively advantageous need in the organism, both in time as well as in their chemical relationships. Enquiry into the nature of the functional relationships of enzymes in present day organisms provides one way of investigating the selective forces that influenced the way in which they evolved. A particularly interesting problem arises with the phosphagen kinases which are inferred to be the products of gene duplication both in this book (Thoai, p. 172) and elsewhere [6, 7]. This group of homologous enzymes appears to be unique in that while the general chemical reaction remains the same

$$\text{guanidine} + \text{ATP} \rightleftarrows \text{phosphoguanidine} + \text{ADP} + \text{H}^+,$$

the specificity requirement for the guanidine has altered such that seven different phosphagens (p. 173) and their kinases are now recognised. The main and perhaps the only function of these enzymes is to produce a rapidly replenishable energy store, particularly in connection with muscular contraction. It is not surprising, then, that there has been rigorous selection for the highly specific nucleotide binding site providing a direct link between ATP generation and its utilization by the contracting myofibril. The evolutionary problem lies in the relation between function and the variability of the guanidine binding site.

A common feature of the guanidines is that they are all amino acids or products of their metabolism and cells may contain significant concentrations of guanidines which are not involved in phosphagen synthesis [8]. In some species they may even have a dietary origin [16]. Hence, by and large, their availability in the cell presents little problem. The 'high-energy' features of the phosphagen reside in the phospho-amidine part of the molecule so that selection for differences in other parts of the guanidine is not influenced by simple energetic considerations. (Although the free energy change of the reaction with different guanidines may vary somewhat, the effective equilibrium position of the reaction is always dictated by the cellular drive to convert all free ADP to ATP.) The basis of selection appears to be metabolic and the chemical form of the guanidine used a question of mechanistic expediency.

Among the invertebrates arginine kinase is the most common kinase, its substrate is a simple, unmodified amino acid and it is the only enzyme known to occur as a monomer. It has long been considered to be the most primitive form from which the others could have evolved. Its first role must have been simply to provide the energy store for a general purpose muscle. An early association with muscular activity is revealed by the finding that this enzyme occurs in the flagellae of the protozoan *Tetrahymena pyriformis* but not to a measurable extent in the body [17]. As animals became more complex a requirement for specialization of muscle function arose. In addition to the original, slow, tonic muscle a rapidly contracting phasic muscle would prove advantageous, for example in avoiding or attacking other species. Among the mollusca divergence in muscle function has, in some species, been accompanied by an obvious divergence in the properties of arginine kinase. As shown for *Cardium edule* in fig. 2 the development of a foot capable of bursts of phasic activity is accompanied by the evolution of a dimeric arginine kinase while the siphon, with a predominantly tonic function, retains the monomer enzyme. The adductor muscles are capable of holding the valves closed over long periods or of making them clap vigorously to clean out the mantle cavity and have approximately equal amounts of both the monomer and dimer arginine kinase. Other Eulamellibranchia show similar features which might be considered characteristic of an advanced molluscan group. However, while most molluscs contain only the monomer arginine kinase the dimeric enzyme has also been found in one primitive gastropod *Patella vulgata* [7]. Such parallel evolution suggests that formation of a dimer molecule confers a selective advantage, perhaps because in the muscle cell the enzyme is located on the dimeric myosin molecule to facilitate energy transfer to a rapidly contracting system [7]. Investigations of other molluscs show that evolution of a dimeric arginine kinase is not prerequisite for phasic activity. Similarly, the presence of only a monomer enzyme does not mean the duplication of the arginine kinase gene has not occurred. The presence of more than one monomer arginine kinase in the scallop, *Pecten maximus*, and squid, *Loligo forbesi*, can be inferred from the different sensitivity of the enzyme from different muscles towards iodoacetamide (table 1).

The association of particular forms of arginine kinase with particular

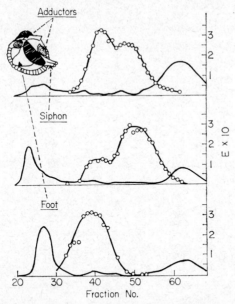

Fig. 2.   Distribution of the 80,000 and 40,000 molecular weight arginine kinases in some muscles of the cockle, *Cardium edule*. The two kinases were separated by chromatography on Sephadex G 100, at pH 7.6, indicated by open circles, with the high molecular weight enzyme peaking near fraction 40 and the low molecular weight enzyme peaking near fraction 50. The protein concentration, solid line, has a peak near fraction 65 due to the inclusion of myoglobin in the sample as an internal molecular weight standard [15].

TABLE 1

Pseudo first order rate constants for the inhibition of some molluscan arginine kinases by iodoacetamide at pH 8.5 and 23 °C

Species	Muscle	$K_I(min^{-1})$
Archaeogastropoda		
Patella vulgata	Foot	0.040
Haliotis tuberculata	Foot	0.055
Pseudolamellibranchia		
Pecten maximus	Smooth adductor	0.109
	Striated adductor	0.152
Eulamellibranchia		
Cardium edule	Adductors	0.039
	Siphon	0.012
	Foot	0.031
Mya truncata	Adductors	0.026
Cephalopoda		
Loligo forbesi	Mantle	0.058
	Tentacle	0.145
	Arm	0.123

types of muscle implies the evolution of appropriate mechanisms regulating the synthesis of each enzyme. This could be one major reason for the occurrence of isoenzymes, when the regulatory machinery of the cell is inadequate to permit a new or modified tissue, which draws largely on the enzyme complement of the parent tissue for its function, to be synthesized alongside the parent tissue. In other words because a new tissue requires a new set of control mechanisms it may also need a new complement of structural genes but not in the same 'all or none' way as a new individual enzyme. Here emphasis is on control rather than just function.

The evolution of new regulatory mechanisms appears to be the underlying reason for the curious distribution of the phosphagen kinases in the Echinodermata (table 2). Species from the different groups contain either arginine kinase, creatine kinase or both and the enzyme distribution is not such as to support the simple view that those animals with both enzymes represent evolutionary intermediates between those with only arginine kinase and those with only creatine kinase. The answer to this problem comes from a study of their gametes. While the eggs from representatives of all classes with the exception of the Crinoidea, which still have to be investigated, contain only arginine kinase the spermatozoa have creatine kinase, sometimes accompanied by arginine kinase [9]. Consequently the fertilized egg must contain the genetic potential to synthesize both arginine and creatine kinases. If the ability to synthesize both enzymes arose after duplication of the arginine kinase gene in a common echinoiderm ancestor, then the origin of the creatine kinase gene must have great antiquity; the enzyme distribution in the

TABLE 2

Distribution of arginine and creatine kinase among the echinodermata*

	Number of species found with		
	Arginine kinase	Creatine kinase	Arginine and creatine-kinase
Crinoidea (sea lilies)	3	0	0
Holothuroidea (sea cucumbers)	13	0	0
Asteroidea (starfish)	14	1	0
Ophiuroidea (brittle stars)	0	9	0
Echinoidea (sea urchins)			
Aulodonta	2	0	0
Stirodonta	1	0	0
Irregularia	4	0	0
Camerodonta	0	1	13

* Data summarized from Moreland et al. [9].

gametes of crinoids, which have an early fossil record, will be of considerable interest in this connection.

What is the nature of the selective process that resulted in the change from arginine to creatine as the phosphagen former in spermatozoa? The most reasonable explanation is that this is associated with the evolution of a highly motile gamete, produced by the testis in large quantities. Such motility necessitates an adequate phosphagen store and would demand a plentiful supply of arginine. At the same time the continued replication of the genome in the testis would require the vast production of arginine-rich histones – a competing interest for the available arginine. Since the phosphagen is in equilibrium with free arginine, in times of amino acid shortage, protein synthesis could cause depletion of the phosphagen store resulting in a non-motile and, consequently, an inviable gamete. The selective advantage of having a phosphagen isolated from the drain of other metabolic processes is obvious. Phosphocreatine fulfills exactly this requirement. Equally important, creatine can be synthesised with virtually no drain on the available arginine supply (fig. 3) by coupling transamidination of the freely-available amino acid, glycine, to the urea cycle as a device for the resynthesis of arginine. The

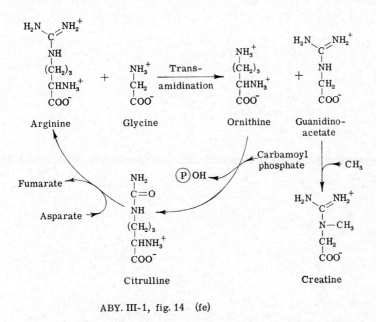

ABY. III-1, fig. 14 (fe)

Fig. 3. Theoretical metabolic pathway for the synthesis of creatine from glycine involving the conservation of arginine by resynthesis from ornithine via the urea cycle [7].

urea cycle, found in both plants and animals, has an antiquity equal to the occasion. It is significant that in mammals the two organs associated with creatine biosynthesis, liver and kidney [10] also contain a functional urea cycle. Similarly, the urea cycle has also been reported to occur in planarians [11] and annelids [12], groups which also have phosphagens other than arginine.

Since the fertilized echinoderm egg contains the genetic potential to synthesize both arginine and creatine kinases, their variable occurrence in the muscles of adults must reflect the further evolution of suitable developmental and control mechanisms. This seems to have occurred independently in the different echinoderm groups so it is not surprising that the distribution of the enzymes in the adults shows no apparent phylogenetic relationships. The question arises as to whether the creatine kinase simply replaced arginine kinase in the otherwise unaltered muscles of these animals or whether a change in muscle function also occurred in a manner similar to that which accompanied dimerisation of arginine kinase in the Mollusca. If it were simply a question of replacement one might expect that those species which contained both kinases would have the same amount of both enzymes in every muscle while functional specialisation, as with the molluscs, would be indicated by the localization of a particular kinase in a particular muscle type. Table 3 shows some data for the kinase distribution in the edible sea urchin, *Echinus esculentus*. The three large well-defined lantern muscles which move the masticatory apparatus contain predominantly arginine kinase while the fine

TABLE 3

The levels of arginine kinase and creatine kinase in some muscles of the edible sea urchin, *Echinus esculentus*

	Enzyme activity (μmoles/mg protein/min)		A/C
	Arginine kinase (A)	Creatine kinase (C)	
Lantern muscles			
Inter-radials	1.0	0.33	3.0
Abductors (teeth retractors)	0.57	0.28	2.0
Adductors (teeth protractors)	0.39	0.36	1.1
Interalveolar (comminators)	1.1	5.05	0.2
Other tissues			
Spine muscle	0.0048	0.096	0.05
Tube feet	0.021	0.175	0.12
Globiferous pedicellaria	0.01	0.083	0.12

comminator muscles which are responsible for the slow action of attrition by the teeth contain much more creatine kinase. One might tentatively describe the two types as being 'more phasic' and 'more tonic' respectively. The tube feet, pedicellaria and spines also have a predominantly tonic function but are capable of bursts of activity under the stimulation of food or danger. These also have mainly creatine kinase. Thus the present evidence supports the idea that in this species, as with the Mollusca the dimeric arginine kinase is associated with phasic function while the less specialized slow muscles have switched to creatine kinase.

The functional association of the two kinases could mean that a new type of tonic muscle has evolved in these species. If this were so then one might expect that the new type of tonic muscle could eventually give rise to a new type of phasic muscle with a dimeric creatine kinase as the active phosphagen former. This is exactly the situation found in modern vertebrates and, gratifyingly, the two muscle types are characterized by different but apparently homologous creatine kinases. One word of caution must be given in that more evidence is required to classify the muscle types in Echinus. The limited evidence available suggests that all the muscles are histologically 'smooth' except for those of the tridentate pedicellaria (not so far examined) which are said to be striated. Such a muscle distribution is clearly not consistent with function and more histological data are required. Nevertheless the biochemical picture which is emerging is consistent with the classical biological concept of evolution [13, 14] that major evolutionary steps always accompany modification of the most primitive biological structure and takes effect from an early stage in development.

As our detailed understanding of the phosphagen kinases grows, so does our appreciation of their function and the way in which they evolved. At the present time the great interest in regulation has made little impact on this group of enzymes which are simply thought to be controlled by the availability and concentrations of their substrates. It could be that a somewhat more sophisticated mechanism operates in some species. Evidence for subunit interaction has been found in human creatine kinase which is not apparent in the rabbit enzyme [18]. The type of evidence is illustrated by fig. 4. Reaction of the two essential thiol groups per enzyme dimer with iodoacetamide is enhanced by a high concentration of $ADP/MgSO_4$, which in the presence of creatine saturates most of the two binding sites per molecule; but is first enhanced and then decreased by a low concentration of the magnesium nucleotide which saturates, on average, only one substrate binding site per molecule. In no case does complete loss of enzyme activity occur. At the present time the results only appear consistent with the hypothesis that alkylation of the thiol at one catalytic site results in a conformational change which makes the thiol on the second catalytic site completely unreactive towards iodoacetamide. Binding of the substrates to the enzyme makes the

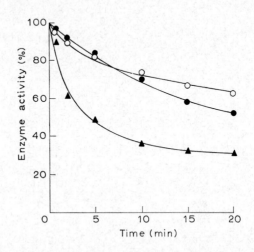

Fig. 4. The effect of creatine plus high and low concentrations of magnesium ADP on the inhibition of human creatine kinase by iodoacetamide. ●, rate of inhibition by iodo-acetamide in the absence of substrates. ○, rate of inhibition in the presence of 8 mM creatine plus 0.25 mM magnesium and ADP. ▲, rate of inhibition in the presence of 8 mM creatine plus 4 mM magnesium and ADP.

thiol at that site more reactive but that at the second site less reactive. In other words a substrate initiated conformational interdependence exists between the two subunits of primate creatine kinase. This type of kinetic behaviour is not seen in the rabbit enzyme where the same substrate combination strongly protects both thiols. Is it possible that divergent evolution has paved the way for allosteric control of the enzyme in one class of animals but not another? A device for enhancing the binding of substrate to the second catalytic site could have a selective advantage to an animal under stress when the phosphocreatine level is low. Whether such a change has come about or merely remains a possibility for the future still has to be determined. It only serves to emphasize the fact that the enzymes we study are themselves evolving and may contain features, the desirability or otherwise of which will only be realized by the selection pressures of tomorrow.

References

[1] R. L. Watts, in Phytochemic phylogeny, J. B. Harborne, ed. (Academic Press, 1969) in press.

[2] R. L. Watts and D. C. Watts, Nature 217 (1968) 1125.

[3] R. L. Watts and D. C. Watts, J. Theoret. Biol. 20 (1968) 227.

[4] A. M. Hayter and R. Riley, Nature 216 (1967) 1028.

[5] H. Lehmann and R. W. Carrell, Brit. Med. Bull. 25 (1969) 14.

[6] D. C. Watts, in Studies in comparative biochemistry, K. A. Munday, ed. (Macmillan, New York, 1965) p. 165.

[7] D. C. Watts, Advan. Comp. Physiol. Biochem. 3 (1968) 1.

[8] N. v. Thoai and J. Roche, in Taxonomic biochemistry and serology, C. A. Leone, ed. (Ronald Press Co., New York, 1964) p. 347.

[9] B. Moreland, D. C. Watts and R. Virden, Nature 214 (1967) 458.

[10] T. R. Koszalka and A. N. Bauman, Nature 212 (1966) 691.

[11] J. W. Campbell, Nature 208 (1965) 1299.

[12] S. H. Bishop and J. W. Campbell, Comp. Biochem. Physiol. 15 (1965) 51.

[13] W. Garstang, Zool. Anz. 17 (1894) 119.

[14] G. R. De Beer, Embryos and ancestors (Oxford University Press, 1951).

[15] B. Moreland and D. C. Watts, Nature 215 (1967) 1092.

[16] G. C. Stephens, J. F. van Pilsum and D. Taylor, Biol. Bull. 129 (1965) 573.

[17] D. C. Watts and L. H. Bannister, Nature 226 (1970) 450.

[18] I. Kumudavalli, B. H. Moreland and D. C. Watts, Biochem. J. 117 (1970) 513.

CAROTENOIDS AND EVOLUTION

T. W. GOODWIN

Department of Biochemistry, The University of Liverpool, England

Carotenoids represent one group of the vast class of natural products known as terpenoids, but they are probably more widely distributed than any other group. For example, monoterpenes are probably not present in algae, and sterols (triterpenoids) are rarely present in bacteria if at all. Carotenoids (tetraterpenes), however, occur in all photosynthetic tissues, where they are located in the chloroplasts of higher plants and algae and in the chromatophores of photosynthetic bacteria [1].

In higher plants and algae where chlorophylls are found they are always accompanied by carotenoids. Any mutation which results in blocking carotenoid synthesis but which has no effect on chlorophyll synthesis, is lethal in the presence of oxygen and light; the mutants are killed by photodynamic sensitization. Thus carotenoids play an essential part in protecting photosynthetic tissues against this sensitization and without their appearance alongside chlorophylls during evolution, higher plants and algae as they are known today would not exist.

When the necessity for the presence of carotenoids in chloroplasts is appreciated, then major taxonomic differences in their distribution would not be expected. This is borne out in higher plants, whilst in algae, although there are differences, when they are considered from the biochemical point of view they are not as great as would appear at first sight.

Before considering the implications of these differences further, the basic structural features of carotenoids and the basic pattern of carotenoid biosynthesis must be considered.

Carotenoids, being tetraterpenes, contain 40 carbon atoms and are made up of branched C-5 isoprene (ip) units arranged thus: ipipipippipipipi. They can be divided into two major groups: carotenes (hydrocarbons) and xanthophylls (oxygen-containing pigments). The main final carotene products of the biosynthetic sequence are lycopene (I) (acyclic), γ-carotene (II) (monocyclic), α-carotene (III) and β-carotene (IV) (bicyclic).

(I)

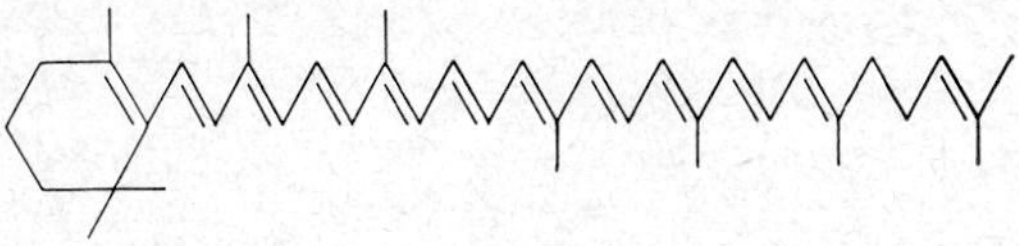

II)

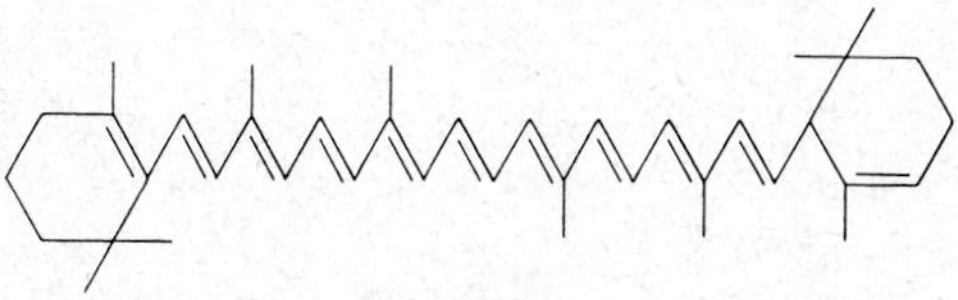

(III)

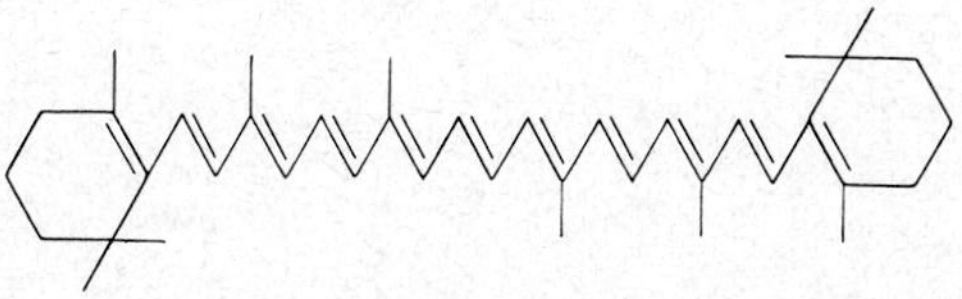

(IV)

The first C-40 compound formed in carotenoid biosynthesis is phytoene which is stepwise desaturated to lycopene (fig. 1). Lycopene or a closely related product is then cyclized to form carotenes with either α- or β-ionone rings. These two ring systems are produced by separate pathways from a common intermediate (fig. 2) and not by isomerization of the formed rings.

The basic xanthophylls formed in green plants are those with hydroxyl groups at positions 3 and 3′, viz. lutein (V;3,3′-dihydroxy-α-carotene and zeaxanthin (VI; 3,3′-dihydroxy-β-carotene).

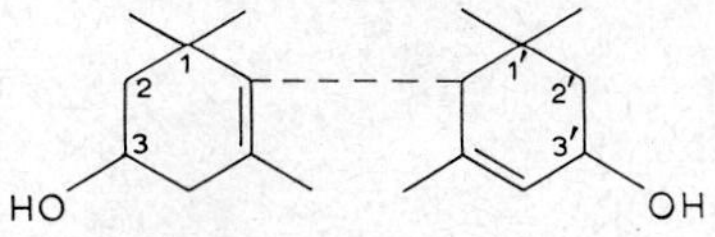

(V) 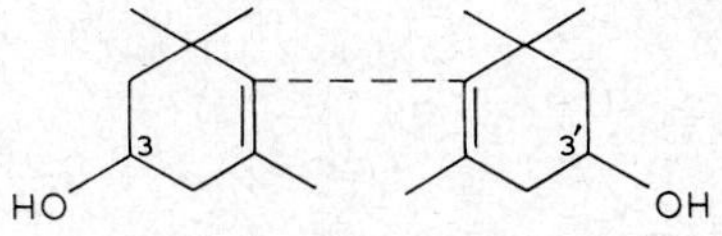   (VI)

In assessing taxonomic or evolutionary significance of carotenoid distribution in photosynthetic tissue any variation in the biosynthetic intermediates listed in fig. 1 cannot be of great significance. Of much greater significance will be the variations which the fully unsaturated carotenoids undergo. These can be related to simple metabolic steps and the variations can thus eb

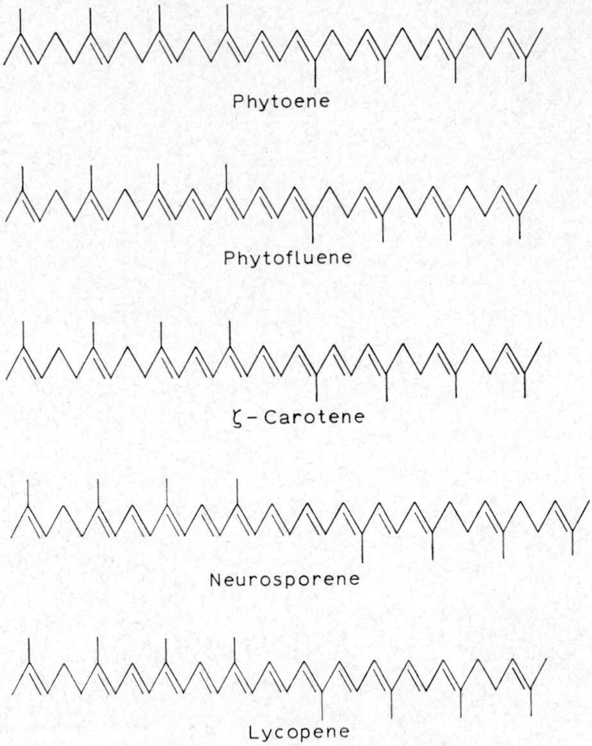

Fig. 1. The desaturation sequence from phytoene to lycopene in higher plants and algae.

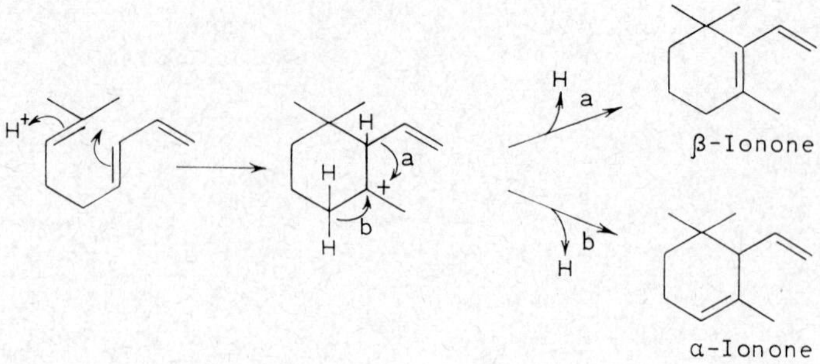

Fig. 2. The mechanism of cyclization to form α-ionone and β-ionone rings in the carotenoids.

TABLE 1

Major pigments of higher plants and associated
specific enzymes

α-Carotene	α-Cyclase	[1]
β-Carotene	β-Cyclase	[2]
Zeaxanthin } Lutein	3-Hydroxylase	[3]
Violaxanthin	Epoxidase	[4]
Neoxanthin	V. isomerase	[5]

attributed to the presence or absence of a specific enzyme, although the enzyme may yet not have been isolated. This concept can be considered first in relation to higher plants, in which the situation is relatively simple, and later in relation to the more complex patterns in algae.

Higher plants. The distribution of the major pigments in higher plants is essentially the same in all species examined. The pigments are listed in table 1. Thus on the basis of the ideas outlined in the previous paragraph, higher plants possess the following "terminal" enzymes (i) α-cyclase (forming α-ionone ring) and β-cyclase (forming β-ionone ring) because they produce both α- and β-carotenes and their derivatives; (ii) 3-hydroxylase (lutein and zeaxanthin); (iii) epoxidase to synthesize violaxanthin (VII; 5,6,5′,6′-diepoxy-zeaxanthin), and (iv) violaxanthin isomerase. The structure of neoxanthin has only recently been elucidated (VIII). It contains an allene group and can reasonably be considered to arise by isomerization of violaxanthin [eq. (1)] under the influence of a postulated enzyme violaxanthin isomerase.

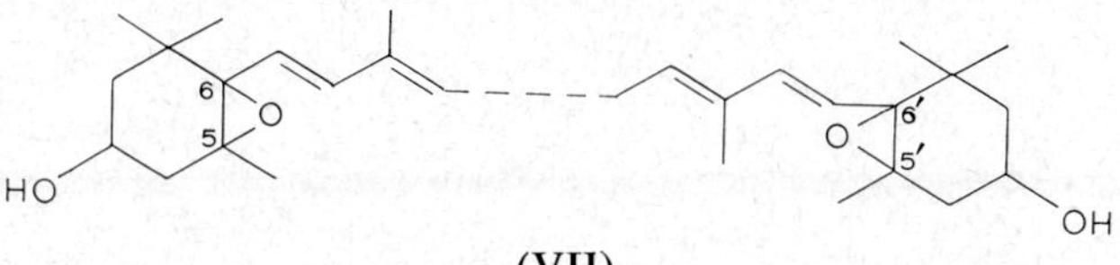

(VII)

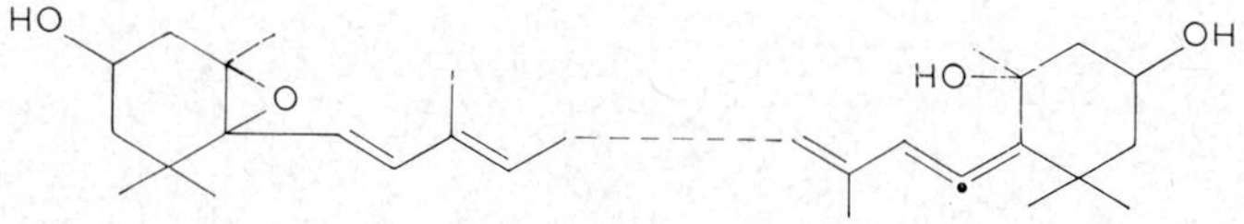

(VIII)

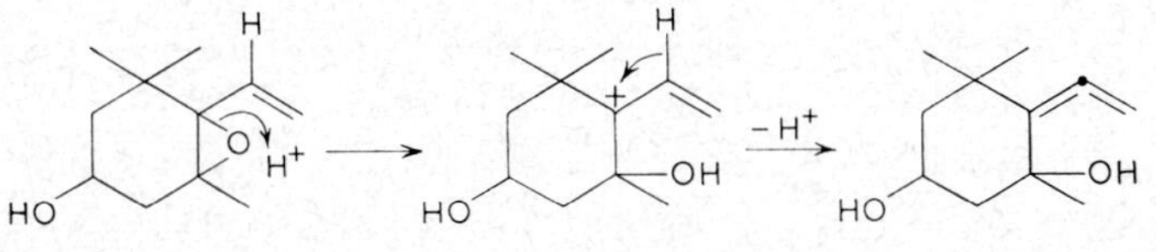

(1)

The general pattern in higher plants (table 1) is observed in plants from all parts of the world, although considerable quantitative variations can be found which may or may not be environmental. Strain who has probably examined more green plants than anyone else concludes that "the chances of finding significant modifications of the pigment system belonging to the groups investigated so far must be very small indeed" [2].

Algae. The situation in algae is more complex than in higher plants but if the approach just discussed is applied the situation becomes clearer and a taxonomic and evolutionary scheme can be constructed (fig. 3) which may form a basis for useful discussion with professional taxonomists.

If one assumes that the primitive precursor of all algae can synthesize only acyclic carotenoids then the first class to emerge is the blue-green algae which

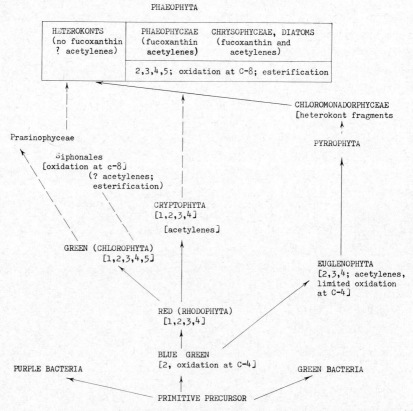

Fig. 3.   An evolutionary tree for photosynthetic bacteria and algae based on carotenoid distribution (numbers refer to enzymes in Table 1).

contains β-cyclase but no α-cyclase; that is only β-carotene and its derivatives are present. In addition these algae have the somewhat uncommon ability of inserting an oxo group into C-4 [echinenone (=myxoxanthin =aphanin) (4-oxo-β-carotene) and canthaxanthin (4,4'-dioxo-β-carotene) are the major xanthophylls] [1, 2]. The red algae could have evolved from the blue-greens by developing an α-cyclase (α-carotene and lutein are present) and a hydro-xylase (zeaxanthin and lutein). In so doing they must have lost the enzyme capable of inserting oxygen into C-4; alternatively the blue-green algae may have evolved this enzyme after the red algae had branched off. Xanthophyll epoxides and neoxanthin are only rarely encountered in red algae [3, 4] so the Chlorophyta, with the exception of the Siphonales, could have evolved from the Rhodophyta by elaborating xanthophyll epoxidase and violaxanthin isomerase and thus producing a pigment system very similar to that of higher plants. The Siphonales were apparently an anomalous group [3] which synthesized in addition to the normal pigments siphonein and siphonaxanthin [5]. The recent elucidation of the structure of siphonaxan-hin (IX) [6], the major pigment, suggests that it can easily be derived from

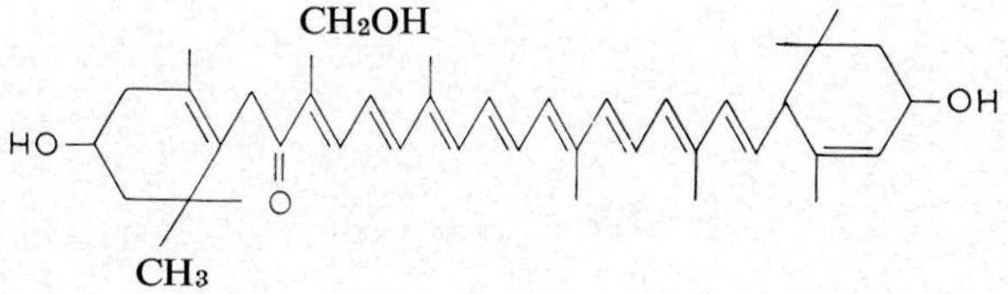

(IX)

chlorophyta pigments by a number of simple metabolic steps: (i) dehydro-genation to an acetylene derivative, hydration of the triple bond to form an enol which could spontaneously assume the keto form [eq. (2)]; (ii) oxidation at C-18, followed by esterification to give siphonein. The last reaction is

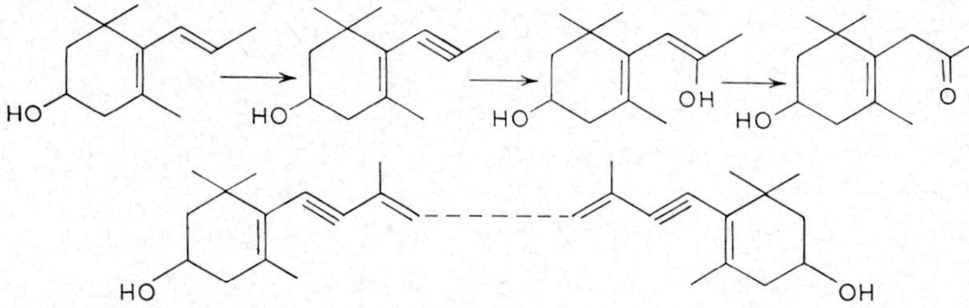

(2)

(X)

interesting in that most xanthophylls in the photosynthetic tissues of higher plants and algae are not esterified. The other major exception is fucoxanthin which has a number of structural affinities with siphonein. This is discussed later. If the mechanism outlined in equation (2) is correct, the traces of acetylenic carotenoids should eventually be detected in the Siphonales. One group of the Prasinophycea (e.g. *Micromonas* spp.) contains a pigment micromone which also has similarities with siphonaxanthin [7]. Others contain siphonein and a third group contains the conventional Chlorophyta carotenoids.

The Cryptophyta could have evolved from the Rhodophyta by evolving an enzyme which produces an acetylenic linkage. Alloxanthin (X) from *Cryptomonas* spp. was the first acetylenic carotenoid to be described [8]. The mechanism involved is not known but the simplest possibility is the direct dehydrogenation of the appropriate double bond.

In the Phaeophyta, the Phaeophyceae, Crysophyceae and Bacillariophyceae synthesize fucoxanthin (XI), the keto function of which could arise from the acetylenic diatoxanthin (XII) by the mechanism suggested for siphonein. Diatoxanthin is present in the Chrysophaceae and Bacillariophyceae, and some acetylenes have been detected in the Phaeophyceae.

The Heterokontae are the only members of the Phaeophyta which do not synthesize fucoxanthin, but *Vaucheria* synthesizes a pigment with acetylenic linkages, a secondary hydroxyl group, a tertiary hydroxyl group and a

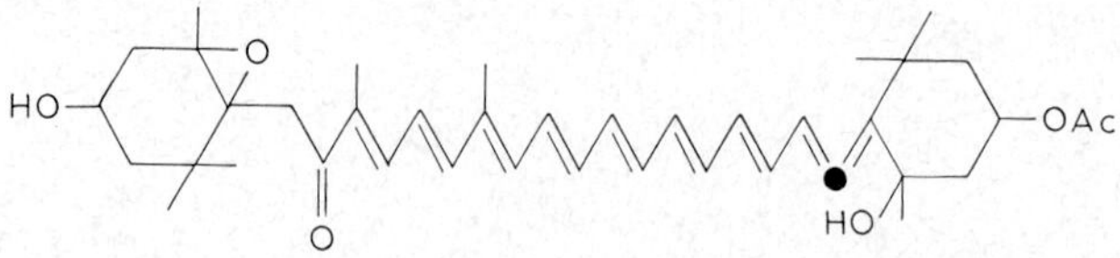

(XI)

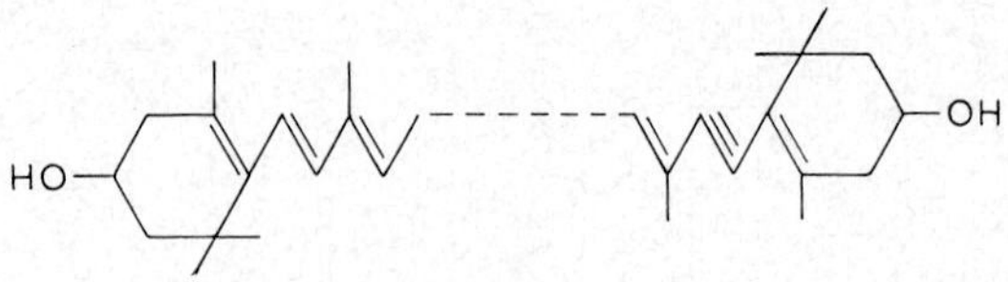

(XII)

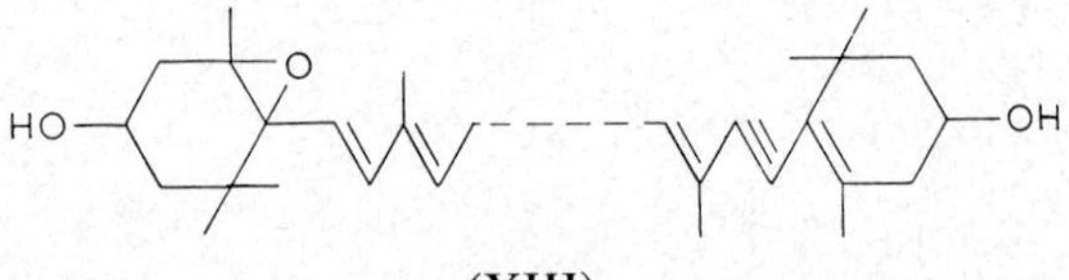

(XIII)

carbonyl group and is presumably of the fucoxanthin/siphonein type.

As the Phaeophyta appear to show little if any α-cyclase activity, it can be suggested that they arise not from the Cryptomonads as previously proposed [1] but by a parallel pathway from blue-green algae via the Euglenophyta and Chloromanadophyceae. *Euglena* spp. synthesize only β-carotene derivatives and its major pigment is the acetylenic diadinoxanthin (XII) [9]. Traces of echinenone, the characteristic blue-green xanthophyll are also present. The Chloromonadophyceae examined so far have the same pigments as the heterokont *Tribonema* [10]. The pigments of the Pyrrophyta are also probably related to those of the Phaeophyceae, e.g. peridinin may be related to fucoxanthin [1], but further modern investigations are required.

Photosynthetic bacteria. Little space is available here to discuss these in detail but in relation to algal evolution, fig. 3 shows that both the purple and the green bacteria can be best considered to have evolved along separate pathways from the primitive precursor which also gives rise to the blue-green algae. With one exception, *Rhodobacterium vanneilii*; the purple bacteria do not contain cyclizing enzymes, but their major 'terminal' activities are hydration of the terminal double bond, methylation of the tertiary alcohol formed and further desaturation at C-3 [eq. (3)]. Other reactions which cannot be considered here also take place [11]. The unique reaction of green bacteria is to

$$(3)$$

carry out a cyclization which does not result in the formation of an α- or β-ionone ring but an aromatic ring as in chlorobactene (XIV) [12]. In some green bacteria γ-carotene is found in traces and may be an intermediate in the formation of an aromatic ring.

(XIV)

Conclusion. In considering carotenoids of photosynthetic tissues in relation to taxonomy both the mandatory requirement for the presence of carotenoids and the basic pattern of biosynthesis must be borne in mind. From these considerations the significant changes to be looked for are variations occurring towards the end of the biosynthetic pathway, i.e. from lycopene onwards. With the newer knowledge of carotenoid structure most of these variations can be reasonably attributed to single enzyme changes. Thus demonstration of a particular 'end product' may give more useful information to the taxonomist and the evolutionist than quantitative studies on pigments and detection or otherwise of normal biosynthetic intermediates.

References

[1] T. W. Goodwin, in Chemistry and biochemistry of plant pigments, T. W. Goodwin, ed. (Academic Press, London, 1964).

[2] H. H. Strain, in Biochemistry of chloroplasts, T. W. Goodwin, ed., Vol. 1 (Academic Press, London, 1966).

[3] H. H. Strain, 32nd Annual Priestly Lecture. Penn. State Univ. (1958).

[4] M. S. Aihara and H. Y. Yamamoto, Phytochemistry (1968).

[5] H. H. Strain, Biol. Bull., 129 (1965) 366.

[6] T. J. Walton, G. Britton, T. W. Goodwin, B. Diner and S. Moshier, Phytochemistry (1970) in press.

[7] T. R. Ricketts, Phytochemistry 6 (1967) 1375.

[8] A. K. Mallams, E. S. Waight, B. C. L. Weedon, D. J. Chapman, F. T. Haxo, T. W. Goodwin and D. M. Thomas, Proc. Chem. Soc. (1967) 301.

[9] K. Aizetmüller, W. A. Svec, J. J. Katz and H. H. Strain, Chem. Commun. 32 (1968).

[10] D. J. Chapman and F. T. Haxo, J. Phycol. 2 (1966) 89.

[11] S. Liaaen-Jensen, in Bacterial photosynthesis, H. Guest, A. San Pietro and L. P. Vernon, eds. (Antioch Press, Yellow Springs, Ohio, 1963) p. 19.

[12] S. Liaaen-Jensen, in Biochemistry of chloroplasts, T. W. Goodwin, ed., Vol. 1 (Academic Press, London, 1966) p. 437.

THE EVOLUTION OF PHOTOCHEMICAL ELECTRON
TRANSFER SYSTEMS

A. A. KRASNOVSKY

A.N. Bakh Institute of Biochemistry, USSR Academy of Sciences, Moscow, U.S.S.R.

Life is sustained on our planet by the photosynthetic energy conversion of light photons. The photosynthetic electron transfer underlies this process of energy conversion.

A comparative biochemical approach facilitates the study of photosynthesis in contemporary photoautotropic organisms among which are the anaerobic photosynthetic bacteria. All these organisms possess an elaborate electron transfer chain composed of pigment systems and biocatalysts arranged in lamellar structures.

Truly primitive photoautotropic organisms no longer exist, and the scientist must turn to other sources for information such as paleontology, and the study of models for abiotic molecular evolution. The use of photochemical studies makes it possible to develop theoretical stages of evolution in photochemical electron transfer. Photoreceptors that absorb and transform light energy are necessary components of any system under study. Such might be inorganic constituents in the earth's crust.

Inorganic photoreceptors. Some metal oxides possess photosensitizing activity among which are the titanium and zinc oxides which absorb in the near ultraviolet. We have shown that it is possible to create an inorganic model of the Hill reaction using these inorganic photoreceptors [1]. These oxides in the presence of near ultraviolet light cause the reduction of ferric ion and the evolution of oxygen from water (fig. 1).

$$2Fe^{+3} + H_2O \rightarrow 2Fe^{+2} + 2H^+ + \tfrac{1}{2}O_2.$$

Such evolution of oxygen has a quantum yield near 1%. No organisms now extant use pure inorganic photoreceptors. For this reason, we have turned to the study of abiogenic organic photosensitizers.

Abiogenic synthesis of pigments. Some information exists on the formation of pigments in model systems of a primitive reducing atmosphere; porphyrins were found in Ponnamperuma's laboratory [2]. Some years ago we studied the conditions of the Rothemund synthesis of porphin in a mixture of formaldehyde and pyrrole using fluorescence spectral measurements as a sensitive method of detection [3], (fig. 2). In the absence of oxygen, the

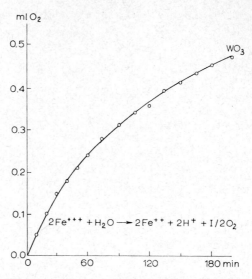

Fig. 1. WO_3 sensitized oxygen evolution. 10 ml H_2O, 0.2 g WO_3, 0.05 $FeNH_4(SO_4)_2$, light intensity (365–436 nm) 2×10^5 erg/cm$^2 \cdot$sec.

reaction proceeds slowly and, in addition to porphin, more reduced pigments are produced – chlorin and probably bacteriochlorin. In air, porphin formation proceeds more rapidly (table 1). In the presence of silica with alumina or titanium dioxide, the porphin fluorescence occurs very quickly. The zinc complexes of porphin are probably formed in the presence of zinc oxide. The red fluorescence sometimes appears if the reaction mixtures are allowed to stand at room temperature.

Porphin isolated from the reaction mixture is an active photosensitizer. We showed that porphin catalyzed the reduction of methyl red by ferrous ion (fig. 3). Ferrous ions were probably abundant as electron donors in the primitive ocean. However, no free porphyrins or inorganic pigments are used as photosensitizers in contemporary organisms possibly because the magnesium porphyrins were the more efficient. The possible explanation of this choice in molecular evolution may be presented in the following deduction [4]. The most abundant metals in the earth's crust are: Si, Al, Fe, Ca, K, Na, Mg, and Ti. There are no silicon, aluminum, or titanium porphyrin complexes in present day organisms probably because of their limited solubilities. The calcium, potassium, and sodium complexes of porphyrins are too easily hydrolyzed in water. Therefore, the iron and magnesium complexes may have been selected out in evolution. Also the magnesium atom in the

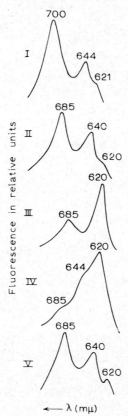

Fig. 2. Fluorescence spectra of porphyrins formed in the mixture of pyrrole and formaldehyde. I: in methanol, vacuo; II: in methanol, aerobic; III: 50% aqueous methanol, O_2; IV: as III + ZnO; V: methanol, SiO_2, O_2.

TABLE 1

The porphyrin formation in the mixture of pyrrole and formaldehyde in 50% aqueous methanol

Condition of experiment	Appearance of red fluorescence (min)
$-O_2$	200
$+O_2$	30
$+O_2 + SiO_2$	15

middle of the porphyrin ring has ligand bonds directed above and below the plane of the ring giving additional bonding capabilities. Iron complexes are non-fluorescent and photochemically inactive, but are able to catalyze electron transfers with an increase in negative free energy in the reactions.

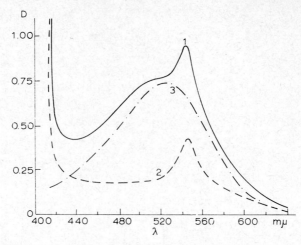

Fig. 3. Porphin photosensitized methyl red reduction by ferrous ions in acid water solution. (1) initial absorption spectra; (2) after illumination; (3) difference spectra 1 and 2 corresponds to methyl red absorption.

Conversely, the magnesium complexes are catalytically inactive, fluorescent and active photochemically. The photosensitizing activity of the magnesium pigments underlines the role of these excited molecules in cyclic oxidation-reduction systems, with their ability to give up electrons.

The primitive organisms probably had both catalytic and photochemical modes of substrate activation. However, where chemically activated substrates of abiogenic origin were abundant, probably there was no need for the latter. Therefore, primitive organisms were probably heterotrophic in accordance with Oparin's hypothesis. The iron complexes were likely used as catalysts along with the flavin and pyridine nucleotides.

The exhaustion of active substrates led inevitably to the use of photochemistry, and the magnesium complexes were incorporated into the primitive catalytic chain. There would have arisen a new type of system – a coupling of photosensitive pigments with iron porphyrin biocatalysts. Such systems are widely distributed in contemporary photoautotrophic organisms. We have prepared models of this primitive system using chlorophyll or its analogs as the photosensitizer and cytochromes, flavins or pyridine nucleotides as catalysts. When chlorophyll was introduced into deoxygenated water solutions of cytochrome c, and the system was illuminated with red light, we observed the photosensitized reduction of the cytochrome. Conversely, the reduced cytochrome underwent photosensitized oxidation in the presence of oxygen [5], (see figs. 4, 5, 6). Also, pyridine nucleotides, ribo-

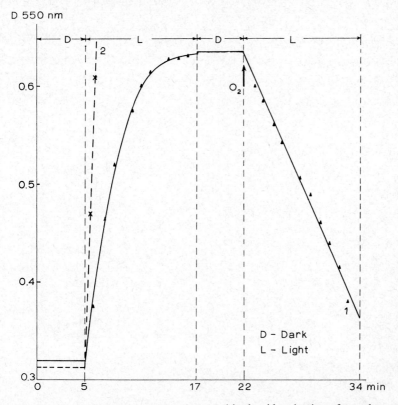

Fig. 4. Kinetics of hematoporphyrin photosensitized oxidoreduction of cytochrome c in 1% aqueous solution of triton X-100 (1), (2) in 50% glycerol.

flavin, and safranin in the presence of ascorbic acid undergo photosensitized reduction by chlorophyll under anaerobic conditions. The reduced pyridine nucleotides are oxidized under aerobic conditions by chlorophyll in the presence of red light.

In summary: In reducing media, chlorophyll and analogs bring about the photosensitized reduction of various electron acceptors, and in oxidizing media, the oxidation of electron donors (review [6]). From these coupled models (pigment-biocatalyst), it is possible to reconstruct different types of electron-transfer chains. The orientation of the pigments in the primitive lipoprotein membrane required an improvement of pigment structure and the lipophylic phytol or farnesol tail gradually became a moiety of the pigment molecule. In the course of evolution, pigment biosynthesis was greatly improved to make better use of the solar energy since a large quantity of pigment had to be present. The accumulation of pigment in membranes led to the phenomena of pigment aggregation. The photosynthetic pigments exist in highly aggregated quasi-crystalline forms in contemporary photo-autotrophs. The phenomenon of light energy migration to pigments directly involved in electron transfer systems was improved.

The evolution of effective photosynthetic organisms was preceded by this

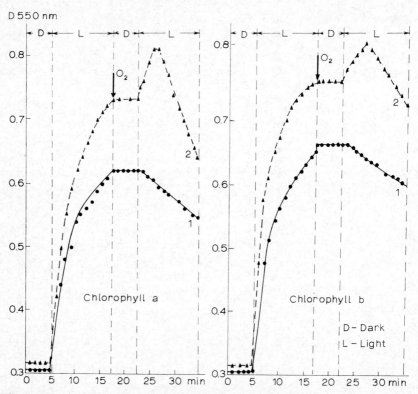

Fig. 5.   Kinetics of chlorophyll photosensitized oxidoreduction of cytochrome c in 1% aqueous solution of triton X-100.

sequence of proper combinations. We proposed in 1957 [4] the gradual evolution from "one quantum" to "two quantum" types of electron transfer. We can propose the following electron transfer chains under anaerobic conditions: one quantum noncyclic transfer from electron donors, i.e., H_2S, organic substances, to an electron acceptor with the formation of active substances such as reduced NADP or ferredoxin; one quantum cyclic electron

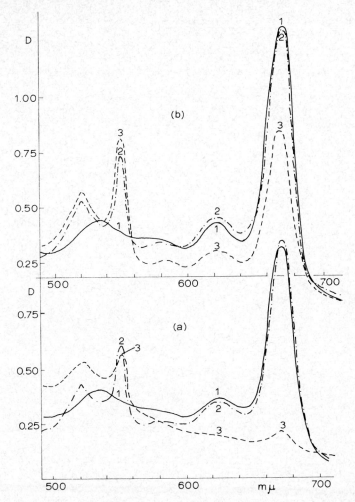

Fig. 6. Absorption spectra changes during chlorophyll sensitized oxidoreduction of cytochrome c. (1) initial spectrum; (2) after illumination in vacuo; (3) after illumination in air.

transfer which led to the formation of ATP from ADP and inorganic phosphate. Combinations of these types probably take place in various photosynthetic bacteria.

The use of water as the ultimate electron donor required a more elaborate system of electron transfer where two photocatalytic systems were involved in electron transfer such as the present I and II photosystems.

References

[1] A. A. Krasnovsky and G. P. Brin, Dokl. Acad. Nauk USSR *168* (1966) 1100; *147* (1962) 655.
[2] G. W. Hodgson and C. Ponnamperuma, Proc. Natl. Acad. Sci. *59* (1968) 22.
[3] A. A. Krasnovsky, and A. V. Umrikhina, Dokl. Acad. Nauk USSR *155* (1964) 691.
[4] A. A. Krasnovsky, in The origin of life on the earth (Pergamon Press, 1959), p. 606.
[5] A. A. Krasnovsky and K. K. Voinovskaia, Biophysica *1* (1956) 120.
[6] A. A. Krasnovsky, Ann. Rev. Plant Physiol. *11* (1960) 363.

THE PRINCIPLE OF EXCESS IN THE
SYNTHESIS OF SECRETIN

MIKLOS BODANSZKY

*Department of Chemistry, Case Western Reserve University,
Cleveland, Ohio, U.S.A.*

Peptide synthesis made impressive advances in the last fifteen years. This productive period started with the synthesis of oxytocin (9 amino acids) by Du Vigneaud and his associates [1] and culminated in the synthesis of β-corticotropin (39 amino acids) by Schwyzer and Sieber [2]. The obvious question whether the synthetic procedures, methods of protection and activation are applicable to the synthesis of proteins with one hundred or more residues still remains to be answered. In this paper only one aspect of this complex problem is brought into consideration: the influence of the concentration of the reactants on the yield and on the homogeneity of the products.

With longer and longer peptide chains there is a corresponding increase in the molecular weights of the synthetic intermediates and hence even if the percent-concentration of the reactants can be kept unchanged from step to step, their molar concentration will decrease. In bimolecular reactions such as the coupling of two peptides, with lessening molar concentration, the number of favorable encounters between the acylating agent (carboxyl component) and the amino component rapidly diminishes. At the same time, side reactions, which are often intramolecular and therefore independent of concentration (e.g., Curtius rearrangement of protected peptide azides), can proceed with an unaffected rate. Thus, a side reaction which is negligible in the coupling of smaller peptides can lead to major amounts of an undesirable product when larger fragments are condensed by the same procedure. Obviously the chances of obtaining a single product are gradually reduced as synthesis proceeds toward longer and longer chains. This situation may present a real limitation in peptide synthesis, especially if the strategy of fragment condensation [3] is applied. In the entirely stepwise approach [4] this basic difficulty can be overcome by the use of a significant excess of acylating agent [5], usually the active ester of a protected amino acid. With an appropriately large excess the reaction becomes pseudo-unimolecular and the desired acylation proceeds with a practical rate even if the amino component is present in low concentration. Removal of the excess acylating agent should not present a major problem since the solubilities in organic solvents of active esters of protected amino acids are quite different from those of protected peptides, especially if the chain contains more than a few amino acids.

These considerations had to be tested in the synthesis of a peptide chain of considerable length. The isolation [6] and elucidation of structure [7] of porcine secretin by Jorpes and Mutt provided the long awaited opportunity for a comparison of different approaches and for the examination of the practicality of the principle of excess. The single chain sequence of porcine secretin containing 27 amino acids is shown in fig. 1. For its synthesis both

```
His-Ser-Asp-Gly-Thr-Phe-Thr-Ser-Glu-Leu-Ser-Arg-Leu-Arg-
 1   2   3   4   5   6   7   8   9   10  11  12  13  14

Asp-Ser-Ala-Arg-Leu-Gln-Arg-Leu-Leu-Gln-Gly-Leu-Val-NH2
 15  16  17  18  19  20  21  22  23  24  25  26  27
```

Fig. 1.   Porcine secretin.

the strategy of fragment condensation and the entirely stepwise approach were planned, the latter with two different techniques: (a) stepwise synthesis through isolated intermediates [4] and (b) synthesis on the polymeric support devised by Merrifield [8]. The last mentioned approach was carried out with the modification proposed by Bodanszky and Sheehan [9] in which protected amino acid nitrophenyl esters [10] are used for acylation instead of coupling with the aid of dicyclohexylcarbodiimide as recommended by Merrifield [8]. This modified approach allows the monitoring of the acylation by examination of the UV spectra of the filtrates, is free from the risks of O-acylation and of dehydration of glutamine and asparagine residues, and can be used also in cases where the presence of free side chain carboxyl groups is desirable. The successive acylations were executed with quite satisfactory results but the removal of the peptide chain from the resin by ammonolysis [11] led to difficulties. When in an exploratory experiment an attempt was made to remove benzyloxycarbonyl-L-valine in the form of its amide by ammonolysis, a mixture consisting of about 20% of the desired amide and ca. 80% of benzyloxycarbonyl-L-valine methyl ester was obtained. This difficulty must be due to steric hindrance by the bulky isopropyl side chain of valine further enhanced by the copolymer. Lengthening of the peptide chain seemed to increase the blocking of the ester bond from a new side. After the C terminal hexapeptide sequence (S_{22-27}) of secretin had been assembled, an attempt was made to remove the chain from the resin. In this case, ammonolysis gave only the protected hexapeptide methyl ester and no amide. Since the methyl esters could be converted into the desired amide by prolonged ammonolysis, the difficulty was not considered to be insurmountable. However, after a protected tridecapeptide corresponding to sequence S_{15-27} had been built up on the solid support, ammonolysis completely failed to remove the peptide. The presence of the protected tridecapeptide on the resin could be ascertained by removal with hydrobromic acid. Thus the solid phase technique in its presently practiced form seemed to be inapplicable for a peptide ending with valine amide and the attempt for the synthesis of secretin on a polymeric support had to be abandoned.

The entirely stepwise synthesis of secretin through isolated intermediates is shown schematically in fig. 2. This approach [12, 13] led to a crude product which exhibited the hormonal properties of natural secretin and had about one-third to one-half of the potency of the hormone. After purification by countercurrent distribution, the synthetic heptacosapeptide was indistinguishable from natural porcine secretin. The comparison included physical data such as distribution coefficients, electrophoretic and paper chromatographic mobilities, the characteristic ORD and CD spectra [14] and a battery of biological tests. Final evidence for the identity of the synthetic and natural products was obtained through side by side digestion of samples with thrombin, trypsin and chymotrypsin and comparison of the resulting fragments. The major impurity in the crude preparations seems to be a peptide in which the aspartyl residue in position 3 is replaced by an aminosuccinyl moiety. No evidence was found for the presence of diastereoisomers that could have been formed by racemization of some amino acid residues: the purified material was fully digestible with proteolytic enzymes specific for peptides of L-amino acids.

The acylating agents, protected amino acid active esters, mainly p-nitrophenyl esters [10] were applied initially in slight, about 10–20%, excess. This excess was gradually increased in such a way that at the onset of the acylation steps the active esters should be present in at least 0.1 molar concentration. To maintain such a concentration it is necessary to increase the amount of the acylating reagent and at the end of the chain lengthening procedure a threefold excess became necessary. The time required for the completion of an acylation reaction, as indicated by the disappearance of the ninhydrin positive amino component from the reaction mixture, was a few hours. In order to allow the acylation of the last and perhaps undetected small amounts of the free amine, the mixtures were kept at room temperature overnight before isolation of the product. This isolation was carried out simply by dilution with ethyl acetate. The excess active ester and the p-nitrophenol

Val-NH$_2$

Leu-Val-NH$_2$

Gly-Leu-Val-NH$_2$

Gln-Gly-Leu-Val-NH$_2$

Leu-Gln-Gly-Leu-Val-NH$_2$

Leu-Leu-Gln-Gly-Leu-Val-NH$_2$

Arg-Leu-Leu-Gln-Gly-Leu-Val-NH$_2$

Gln-Arg-Leu-Leu-Gln-Gly-Leu-Val-NH$_2$

Leu-Gln-Arg-Leu-Leu-Gln-Gly-Leu-Val-NH$_2$

Arg-Leu-Gln-Arg-Leu-Leu-Gln-Gly-Leu-Val-NH$_2$

Ala-Arg-Leu-Gln-Arg-Leu-Leu-Gln-Gly-Leu-Val-NH$_2$

Ser-Ala-Arg-Leu-Gln-Arg-Leu-Leu-Gln-Gly-Leu-Val-NH$_2$

Asp-Ser-Ala-Arg-Leu-Gln-Arg-Leu-Leu-Gln-Gly-Leu-Val-NH$_2$

Arg-Asp-Ser-Ala-Arg-Leu-Gln-Arg-Leu-Leu-Gln-Gly-Leu-Val-NH$_2$

Leu-Arg-Asp-Ser-Ala-Arg-Leu-Gln-Arg-Leu-Leu-Gln-Gly-Leu-Val-NH$_2$

Arg-Leu-Arg-Asp-Ser-Ala-Arg-Leu-Gln-Arg-Leu-Leu-Gln-Gly-Leu-Val-NH$_2$

Ser-Arg-Leu-Arg-Asp-Ser-Ala-Arg-Leu-Gln-Arg-Leu-Leu-Gln-Gly-Leu-Val-NH$_2$

Leu-Ser-Arg-Leu-Arg-Asp-Ser-Ala-Arg-Leu-Gln-Arg-Leu-Leu-Gln-Gly-Leu-Val-NH$_2$

Glu-Leu-Ser-Arg-Leu-Arg-Asp-Ser-Ala-Arg-Leu-Gln-Arg-Leu-Leu-Gln-Gly-Leu-Val-NH$_2$

Ser-Glu-Leu-Ser-Arg-Leu-Arg-Asp-Ser-Ala-Arg-Leu-Gln-Arg-Leu-Leu-Gln-Gly-Leu-Val-NH$_2$

Thr-Ser-Glu-Leu-Ser-Arg-Leu-Arg-Asp-Ser-Ala-Arg-Leu-Gln-Arg-Leu-Leu-Gln-Gly-Leu-Val-NH$_2$

Phe-Thr-Ser-Glu-Leu-Ser-Arg-Leu-Arg-Asp-Ser-Ala-Arg-Leu-Gln-Arg-Leu-Leu-Gln-Gly-Leu-Val-NH$_2$

Thr-Phe-Thr-Ser-Glu-Leu-Ser-Arg-Leu-Arg-Asp-Ser-Ala-Arg-Leu-Gln-Arg-Leu-Leu-Gln-Gly-Leu-Val-NH$_2$

Gly-Thr-Phe-Thr-Ser-Glu-Leu-Ser-Arg-Leu-Arg-Asp-Ser-Ala-Arg-Leu-Gln-Arg-Leu-Leu-Gln-Gly-Leu-Val-NH$_2$

Asp-Gly-Thr-Phe-Thr-Ser-Glu-Leu-Ser-Arg-Leu-Arg-Asp-Ser-Ala-Arg-Leu-Gln-Arg-Leu-Leu-Gln-Gly-Leu-Val-NH$_2$

Ser-Asp-Gly-Thr-Phe-Thr-Ser-Glu-Leu-Ser-Arg-Leu-Arg-Asp-Ser-Ala-Arg-Leu-Gln-Arg-Leu-Leu-Gln-Gly-Leu-Val-NH$_2$

His-Ser-Asp-Gly-Thr-Phe-Thr-Ser-Glu-Leu-Ser-Arg-Leu-Arg-Asp-Ser-Ala-Arg-Leu-Gln-Arg-Leu-Leu-Gln-Gly-Leu-Val-NH$_2$

1   2   3   4   5   6   7   8   9   10 11 12 13 14 15 16 17 18 19 20 21 22 23 24 25 26 27

Fig. 2.   The entirely stepwise strategy.

produced are soluble in this solvent and are completely removed when the
insoluble protected peptide is washed with ethyl acetate. The protected
intermediates were obtained as easily tractable solids, most of them crystalline, with satisfactory elemental and amino acid analysis. The average yield

for the lengthening of the chain by one amino acid is 94%. Since this figure does not reflect the heterogeneity mentioned earlier and because of the not insignificant excess on many of the acylating components, the calculation of an "overall yield" for the entire synthesis is probably unwarranted.

The fragment condensation approach [15] is summarized in fig. 3, which

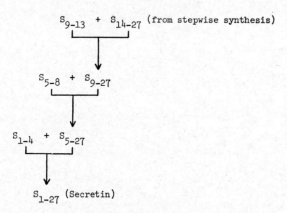

Fig. 3. Synthesis of secretin by fragment condensation.

shows that even in this strategy some of the characteristics of the stepwise approach were maintained: a gradual lengthening of the C terminal sequence and the use of the acylating agents (protected peptide azides) in significant excess.

The first condensation, coupling of (protected) S_{9-13} to S_{14-27}, could be accomplished with a yield of about 80%, calculated on the amino component. The C terminal tetradecapeptide derivative was prepared by the entirely stepwise strategy. The resulting nonadecapeptide S_{9-27} was obtained in pure form simply by extraction with water. The subsequent couplings, S_{5-8} to S_{9-27} and S_{1-4} to S_{5-27}, gave 50 and 33% yields respectively and for isolation of the products in pure form more elaborate procedures, such as counter-current distribution or chromatography, were necessary. Once again the calculation of an "overall yield" should be avoided, because a single figure would not reflect the sacrifices on some of the intermediates (e.g., on S_{9-13}). The product of the fragment condensation approach was shown to be identical both with natural porcine secretin and with the synthetic hormone prepared by the entirely stepwise strategy.

The gradual decrease in the yields in subsequent coupling steps seems to be a rather characteristic feature of peptide syntheses by fragment condensation. One is under the impression that in the synthesis of peptide chains a limit is being approached because the difficulties increase with the length of the chain more than proportionally. In the entirely stepwise strategy, no such deterioration of yields could be observed. The possibility of purification of intermediate products is somewhat better in the case of fragment condensation especially if unprotected amino components are used, since these are more easily purified than the protected peptides that are insoluble in most solvent systems used for chromatography or countercurrent distribution. Also, the impurities are often more different from the desired products in the fragment condensation approach than in the stepwise strategy. This advantage should not be belittled. Nevertheless, since the entirely stepwise strategy [4] allows the systematic application of the principle of excess, and a priori promises less heterogenous products, it seems possible to adopt the stepwise synthesis with active esters also for more ambitious endeavors, for the synthesis of proteins.

References

[1] V. du Vigneaud, C. Ressler, J. M. Swan, C. W. Roberts, P. G. Katsoyannis and S. Gordon, J. Am. Chem. Soc. *75* (1953) 4879.
[2] R. Schwyzer and P. Sieber, Helv. Chim. Acta *49* (1966) 134.
[3] M. Bodanszky and M. A. Ondetti, in Peptide synthesis (Interscience, New York, 1966) Ch. VII.
[4] M. Bodanszky, Ann. N. Y. Acad. Sci. *88* (1960) 655.
[5] M. Bodanszky and A. Bodanszky, Am. Scientist *55* (1967) 185.
[6] J. E. Jorpes and V. Mutt, Acta Chem. Scand. *15* (1961) 1790.
[7] V. Mutt and J. E. Jorpes, presented at the 4th International Symposium on the Chemistry of Natural Products, Stockholm, Sweden, 1966; cf. also Pharmacology of hormonal polypeptides and proteins, N. Back, L. Martini and R. Paoletti, eds. (Plenum Press, New York, 1968) p. 569.
[8] R. B. Merrifield, J. Am. Chem. Soc. *86* (1964) 304.
[9] M. Bodanszky and J. T. Sheehan, Chem. Ind. (1964) 1423.
[10] M. Bodanszky, Nature *175* (1955) 685.
[11] M. Bodanszky and J. T. Sheehan, Chem. Ind. (1966) 1597.
[12] M. Bodanszky and N. J. Williams, J. Am. Chem. Soc. *89* (1967) 685.
[13] M. Bodanszky, M. A. Ondetti, S. D. Levine and N. J. Williams, J. Am. Chem. Soc. *89* (1967) 6753.
[14] A. Bodanszky, M. A. Ondetti, V. Mutt and M. Bodanszky, J. Am. Chem. Soc. *91* (1969) 944.
[15] M. A. Ondetti, V. L. Narayanan, M. von Saltza, J. T. Sheehan, E. F. Sabo and M. Bodanszky, J. Am. Chem. Soc. *90* (1968) 4711.

INVESTIGATION OF A TOTAL SYNTHESIS OF FERREDOXIN; SYNTHESIS OF THE AMINO ACID SEQUENCE OF C. PASTEURIANUM FERREDOXIN*

ERNST BAYER, GÜNTHER JUNG and HANSPAUL HAGENMAIER

Chemisches Institut der Universität Tübingen, Germany and Chemistry Department, University of Houston, Texas, U.S.A.

The total synthesis of ferredoxin, a compound participating in electron transfer during assimilation process, may be divided into three parts:

1. Total synthesis of the amino acid sequence [1] from the 55 amino acids constituting Clostridium ferredoxin.
2. Synthesis of a secondary or tertiary structure suitable for the introduction of the active center.
3. Introduction of the iron-containing active center.

Previous experiments have shown that the third step, i.e., the introduction of the active center into iron-free inactive apoferredoxin by treatment with α,α'-dipyridyl results in a 40% yield of plant ferredoxin [2] and a 70% yield of Clostridium ferredoxin [3]. In both cases, biological activity is completely restored. The high yield obtained during this third and last step encouraged us to attempt total synthesis.

The initial and also the most time-consuming step consists in the coupling

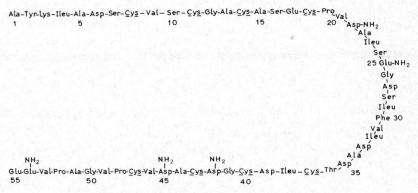

Fig. 1.   Amino acid sequence of *C. pasteurianum* ferredoxin (1).

* Reprinted by permission from Tetrahedron *24* (1968) 4853.

of 55 amino acids to produce the primary structure of *C. Pasteurianum* ferredoxin (fig. 1). The solid phase method of Merrifield [4], by which the working time involved is materially shortened, seemed to be most suitable. However, in contrast to classical methods, it does suffer the disadvantage of forming a large number of difficultly separable peptide mixtures, a situation compounded even further by yields of less than 100% and the statistical distribution of each newly coupled amino acid among the principal different polypeptide chains. Preparation of the amino acid sequence of apoferredoxin should, therefore, provide fundamental evidence as to the feasibility of the Merrifield method, even more so in light of the fact that up to the present time* only the 21–30 amino acid chains of insulin have been synthesized [5]. One further difficulty is presented by the high cysteine content of ferredoxin, as the splitting off of the protecting benzyl group could result in extensive changes in the molecule. We determined by preliminary experiments that natural apoferredoxin, after treatment with sodium in liquid ammonia, can be resynthesized to active ferredoxin in 15% yield. Therefore, the removal of the protecting S-benzyl group would not be a problem in the case of apo-ferredoxin.

From the point of view of obtaining maximum yields during the course of the peptide synthesis, the t-butoxycarbonyl amino acids were first purified in the most exacting manner, the final products being checked by mass spectrometry [6], gas chromatography [6], polarimetry and thin-layer chromatography. Data indicative of the final purity of these t-butoxycarbonyl (BOC) compounds are given in table 1. BOC-amino acids were prepared both by the procedures of Schwyzer et al. [7] and by the pH- stat method of Schnabel [8]. The latter was found to be preferable, in that the BOC-amino acids were formed in better yield and purity.

For the synthesis of sequence (1), the following procedure was utilized [9]. The COOH-terminated glutamic acid (sequence No. 55) was esterified in benzene as the BOC—Glu(OBz)—OH derivative [10] with a chloromethy-lated copolymer of styrene and divinyl benzene (Bio-Rad, New York, capacity: 1.8 mMol/g). The normal procedure of using ethanol instead of benzene was not used in this case due to the possibility of transesterification. The resultant loading on the resin was determined by amino acid analysis following hydrolysis in 6 N HCl and was found to be 0.28 mMol per gram of resin. An alternate method of hydrolysis consisting of refluxing

---

* Since this publication went to press Merrifield and Gutte reported the synthesis of a protein with ribonuclease activity.

TABLE 1

Properties of the t-butoxycarbonyl amino acids used in this work

1. BOC-L-Ala-OH — Mp. 81 °C (Lit. 83–84 °C [1], 80–82 °C [2]); $[\alpha]_D^{22} = -24.9$ ($c = 2.049$, glacial acetic acid) (Lit. $[\alpha]_D^{25} = -22.4°$ ($c = 2.095$, glacial acetic acid)[1]; $[\alpha]_{578}^{18-25} = -25.2°$ ($c = 1$, glacial acetic acid) [2])

2. BOC-L-Asn-OH — Mp. 180 °C (Lit. 181–182 °C [3]), 200 °C [4], 174–176 °C [2]); $[\alpha]_D^{28} = -7.6°$ ($c = 1.097$, DMF) (Lit. $[\alpha]_D^{25} = -7.8°$ ($c = 1$, DMF) [3]; $[\alpha]_{578}^{18-26} = -8.5°$ ($c = 1$, DMF) [2])

3. BOC-L-Asn-OHp — Mp. 157–158 °C (Lit. 157–158 °C [3], 163 °C [5]); $[\alpha]_D^{25} = -40.4°$ ($c = 1.76$, methanol) and $[\alpha]_D^{28} = -36.9°$ ($c = 1.001$, DMF) (Lit. $[\alpha]_D^{25} = -40°$ bzw. $-45.3°$ ($c = 1$, DMF) [3]; $[\alpha]_D^{22} = -35.0°$ ($c = 2$, glacial acetic acid 95 %) and $[\alpha]_D^{22} = -36.0°$ ($c = 2$, DMF) [4])

4. BOC-L-Asp(OBzl)-OH — Mp. 99 °C (Lit. 101 °C [4]); $[\alpha]_D^{27} = +8.2°$ ($c = 2.153$, glacial acetic acid) and $[\alpha]_D^{26} = +36.5°$ ($c = 1.973$, methanol) (Lit. $[\alpha]_D^{22} = +9° \pm 1°$ ($c = 2.95$, glacial acetic acid 95 %) [4])

5. BOC-L-Asp(OBzl)-ONp — Mp. 105 °C; $[\alpha]_D^{26} = -24.5°$ ($c = 2.132$, methanol)

6. BOC-L-Cys(Bzl)-OH — Mp. 86 °C (Lit. 63–65 °C [1]; 86–87 °C [6]; 63–65 °C [2]); $[\alpha]_D^{25} = -43.2°$ ($c = 1.027$, glacial acetic acid) (Lit. $[\alpha]_D = -41.0°$ ($c = 0.999$, glacial acetic acid [1]; $[\alpha]_{578}^{18-25} = -45.3°$ ($c = 1$, glacial acetic acid [2])

7. BOC-L-Gln-OH — Mp. 116–117 °C with decomposition (Lit. 114–118 °C [7], 116–118 °C [2]); $[\alpha]_D^{27} = -4.6°$ ($c = 1.944$, ethanol) (Lit. $[\alpha]_D^{25} = -3.0°$ ($c = 1$, ethanol [7]; $[\alpha]_{578}^{18-25} = -2.9°$ ($c = 1$, ethanol) [2])

8. BOC-L-Gln-ONp — Mp. 156–157 °C (Lit. 145–146 °C [8]); $[\alpha]_D^{25} = -41.1°$ ($c = 1.995$, methanol)

9. BOC-L-Glu-(OBzl)-OH — Colorless oil Öl (Lit. Öl [9]) $[\alpha]_D^{27} = -6.0$ ($c = 1.912$, glacial acetic acid)

10. BOC-Gly-OH — Mp. 88 °C (Lit. 85–89 °C [10], 88.5–89 °C and 89–90 °C [1])

11. BOC-Gly-ONp — Mp. 70 °C (Lit. 70–71 °C [4])

12. BOC-L-Ille-OH × 0.5 H₂O — Mp. 59–61 °C (Lit. 49–57 °C [1], 66–68 °C [2]) $[\alpha]_D^{27} = +2.4°$ ($c = 1.653$, glacial acetic acid) and $[\alpha]_D^{28} = +3.8°$ ($c = 2.080$, methanol) (Lit. $[\alpha]_D^{25} = +3.0°$ ($c = 2.005$, glacial acetic acid [1]); $[\alpha]_{578}^{18-25} = +2.5°$ ($c = 1$, glacial acetic acid) [2])

13. BOC-L-Lyz(Z)-OH — Colorless oil Öl (Lit. Öl [1]) $[\alpha]_D^{27} = -8.2°$ ($c = 2.102$, glacial acetic acid) (Lit. $[\alpha]_{578}^{18-25} = -9.3°$ ($c = 1$, glacial acetic acid) [2])

14. BOC-L-Phe-OH — Mp. 65 °C (Lit. Öl [11]; 79–80 °C [1]; 85–87 °C [12]; 84–86 °C [2]), $[\alpha]_D^{25} = -3.6°$ ($c = 5.048$, glacial acetic acid (Lit. $[\alpha]_D = -0.8°$ ($c = 4.975$, glacial acetic acid) [1]; $[\alpha]_D^{20} = 4.1 \pm 2°$ ($c = 5$,

226   E. BAYER, G. JUNG AND H. HAGENMAIER

Table 1 (Continued)

	glacial acetic acid [12]; $[\alpha]_{578}^{18-25} = -4.0°$ ($c = 1$, glacial acetic acid) [8])
15. BOC-L-Pro-OH	Mp. 133–134°C (Lit. 136–137°C [1]; 134–136°C [1]) $[\alpha]_D^{25} = -60.8°$ ($c = 2.043$, glacial acetic acid) (Lit. $[\alpha]_D = -60.2°$ ($c = 2.011$, glacial acetic acid) [1]; $[\alpha]_{578}^{18-25} = -68.5°$ ($c = 1$, glacial acetic acid) [2]
16. BOC-L-Ser(Bzl)-OH	Mp. 59°C (Lit. 60–62°C [13]; Öl [9]) $[\alpha]_D^{27} = +18.8°$ ($c = 1.598$, ethanol 80%) (Lit. $[\alpha]_D^{20} = +20.4°$ ($c = 2$, ethanol 80%) [13])
17. BOC-L-Thr-OH	Mp. 80°C (Lit. 76–80°C [7]; 74–77°C [2]) $[\alpha]_D^{25} = -2.0°$ ($c = 0.998$, methanol) (Lit. $[\alpha]_D = -2.52°$ (methanol) [7])
18. BOC-L-Tyr(Bzl)-OH	Mp. 109–110°C (Lit. 108–110°C [2]) $[\alpha]_D^{26} = +16.35°$ ($c = 2.077$, methanol) (Lit. $[\alpha]_{578}^{18-25} = +12.1°$ ($c = 1$, acetone) and $+5.5°$ ($c = 1$, glacial acetic acid) [2])
19. BOC-L-Val-OH	Mp. 80°C (Lit. Öl [11]; 72–73°C [2]; 77–79°C [1]) $[\alpha]_D^{28} = -6.9°$ ($c = 2.009$, glacial acetic acid) (Lit. $[\alpha]_D^{25} = 5.8°$ ($c = 1.208$, glacial acetic acid) [1]; $[\alpha]_{578}^{18-25} = +6.0°$ ($c = 1$, glacial acetic acid) [2])

Literature references to table 1

[1] G. W. Anderson and A. C. McGregor, J. Am. Chem. Soc. *79* (1957) 6180.
[2] E. Schnabel, Liebigs Ann. Chem. *702* (1967) 188.
[3] E. Schröder and E. Klieger, Liebigs Ann. Chem. *673* (1964) 208.
[4] E. Sandrin and R. A. Boissonas, Helv. Chim. Acta *46* (1963) 1637.
[5] G. R. Marshall and R. B. Merrifield, Biochemistry *4* (1965) 2394.
[6] R. Geiger, private communication to E. Wünsch, Methoden der organischen Chemie, Houben-Weyl-Müller Bd. XV, in press.
[7] K. Hofmann, R. Schmiechen, R. D. Wells, Y. Wolman and N. Yanaihara, J. Am. Chem. Soc. *87* (1965) 611.
[8] H. Zahn, W. Danho and B. Gutte, Z. Naturforsch. *21* (1966) 673.
[9] R. Schwyzer, B. Iselin, H. Kappeler, W. Rittel and P. Sieber, U.S. Patent 3, 247, 182, Pat. April 19 (1966).
[10] F. C. McKay and N. F. Albertson, J. Am. Chem. Soc. *79* (1957) 4686.
[11] R. Schwyzer, P. Sieber and H. Kappeler, Helv. Chim. Acta *42* (1959) 2622.
[12] E. Wünsch and G. Wendlberger, Chem. Ber. *97* (1964) 2504.
[13] E. Wünsch and A. Zwick, Chem. Ber. *97* (1964) 2497.

in 6 N HCl-dioxane-water was found to be unsuitable, since up to 50% threonine and serine are produced from the glutamic acid.

The removal of the BOC protecting group was accomplished by means of 1 N HCl in glacial acetic acid [4, 5], the time required increasing to 40 min toward the end of the synthesis. This method of BOC-removal is preferred to that using 4 N HCl-dioxane, a reagent applied only in the case of threonine where it was necessary to avoid acylation of the unprotected hydroxyl group. The reaction mixture was neutralized in the normal manner with triethylamine in dimethylformamide, the resulting triethylamine hydrochloride being utilized for the determination of free amino groups available in the next coupling step by Vollhard titration [10]. Fig. 2 illustrates the continuous decrease in the number of free amino groups during the course of this synthesis. After 54 coupling steps, 38% of the free amino groups, as determined following the first BOC removal, are still available. This corresponds to an average yield of 98.9% per synthetic step.

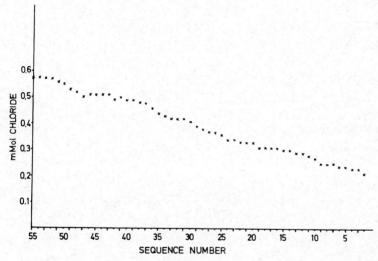

Fig. 2.   Titration of triethylamine hydrochloride during the course of the synthesis of apoferredoxin.

The formation of the peptide linkage was brought about by dicyclohexyl-carbodiimide (DCCI) in methylene chloride. As the chain length of the poly-peptide continued to grow, reaction times increased from two up to three hours and even before addition of the DCCI, a good sorption of the BOC-amino acid solution in the resin was observed. For each synthetic step, 2.5–3 mMol of BOC-amino acid were added per 2 g of resin, amounting to a 5–6 fold excess. Asn, Gln and Gly were used as the pNp-esters in 6–7 fold excess for 6–8 hr reaction time. In addition, the Asp following Thr was introduced as the BOC-Asp (Bzl)-ONP derivative, since the Thr, as was the case in the insulin synthesis, was used with an unprotected hydroxyl group [11]. The course of these reactions was followed by analysis of the hydrolyzate with an automatic amino acid analyzer (Model Unichrom, Beckman Instr., Munich). The analytical results show an increase of the corresponding amino acids in good agreement with expected values.

The protected polypeptide polymer was split off from the resin by passing bromine-free HBr through a suspension of the polypeptide-resin in trifluoro-acetic acid in the presence of methionine. The residual free resin was filtered off and the polypeptide precipitated by addition of anhydrous ether, yielding 1.75 g of S-benzyl polypeptide as a white powder. This corresponds to a yield of 51% based on the amount of C-terminal glutamic acid initially bound to the resin. This increased yield (greater than the 38% value obtained by titration) may be explained by the presence of smaller peptide chains.

In order to split off the protecting S-benzyl groups, 500 mg of the S-benzyl polypeptide were dissolved in 650 ml of liquid ammonia. Sodium, also dissolved in liquid ammonia, was then added dropwise from a cooled dropping funnel until a blue color, stable for 60 sec, indicated completion of the reaction. The bulk of the ammonia was then evaporated to a volume of 30 ml, the remainder being removed by freeze drying under water-aspirator vacuum. The resulting powder was dissolved in 0.2% acetic acid, brought to pH 9.0 by the addition of tris-(hydroxymethyl)-amino methane and then subjected to oxidative sulfitolysis [12] by addition of sodium sulfite and sodium tetrathionate. After 24 hr, the resulting S-sulfonate was desalted by dialysis against water and upon freeze drying yield 450 mg of a light white powder.

The S-sulfonate was purified on a 108×2.5 cm column of Sephadex G-50 using 0.2 N acetic acid as eluent. Extinction of the fractions obtained was recorded automatically at 250 mμ by a Uvicord (LKB-Produkter AB, Stockholm). One peak was a found via the Folin-Lowry test and the Folin-positive fractions were purified twice more over Sephadex G-50. In this case also, a peak was found by both a Folin–Lowry test and by UV absorption whose elution time corresponded to that of the S-sulfonate of natural apoferredoxin (fig. 3). After freeze drying, 103 mg of a white powder were obtained from an initial sample of 500 mg of the S-benzyl polypeptide.

The UV spectrum of the synthetic S-sulfonate, as well as that of the S-sulfonate of natural apoferredoxin, shows a "shoulder" at 280 mμ which may be attributed to tyrosine (fig. 4). The molecular weight, determined by sedimentation equilibrium ultracentrifuge, was found to be 6320 ± 300 (calc. wt. from the sequence: 6134) with an assumed partial volume of 0.71 *. The

* We thank Prof. Dr. H. Friedrich-Freksa and H. Bachowsky of the Max-Planck-Institut für Virusforschung for the ultracentrifuge measurements.

purified polypeptide S-sulfonate itself, is uniform in the ultracentrifuge. The amino acid analysis (table 2) of the synthetic product corresponds to that of natural apoferredoxin. An ammonia determination (calc: 5.0, found: 5.4) shows that the acid amide groups remain intact.

An end-group determination of the N-terminal amino acids was carried out with 2,4-dinitrofluorobenzene [13] and following thin-layer chromato-

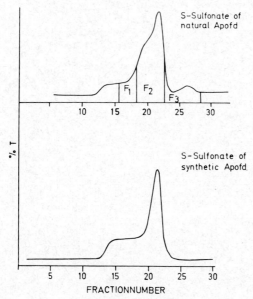

Fig. 3.   Chromatogram of natural and synthetic apoferrodoxin-S-sulfonate on Sephadex G-50.

graphy in two different developing solvents, DNP-alanine was found. A further and weaker spot was observed next to the DNP-alanine and could correspond to either DNP-tyrosine or DNP-lysine. These findings are identical to those encountered during the end-group analysis of natural apoferredoxin.

In light of the results of the end-group determinations, the ultra-centrifuge measurements, the clear-cut course of the synthesis, the analytical data and chromatographic findings, it is clearly established that the amino acid sequence of ferredoxin, as given by Tanaka et al., has been synthesized.

At the same time, the synthesis of the 55 amino acid sequence also shows the Merrifield method to be suitable for the preparation of high molecular weight polypeptides. It is interesting to notice from fig. 3 that both the S-sul-

TABLE 2
Amino acid analyses

Peptide corresponding to sequence no.	48–55		33–55		30–55		17–55		10–55		1–55		I	II	III	IV
Temperature [°C]	118		112		112		112		114		110		For I–IV, see under			
Time (hr)	24		22		21		28		21		26					
Amino acid	E*	T**	E	T	E	T	E	T	E	T	E	T	E	E	E	E
Glu	2.5	2	2.7	2	3.0	2	4.8	4	5.6	4	5.5	4	4.3	4.2	4.3	4.1
Val	2.1	2	3.4	3	3.8	4	5.3	5	5.2	5	7.0	6	6.2	5.5	4.9	4.8
Pro	2.1	2	2.7	2	2.1	2	3.3	3	4.5	3	3.6	3	4.2	3.2	2.8	3.1
Ala	1.1	1	2.4	2	3.3	3	4.6	4	5.7	6	6.6	8	6.9	6.7	7.0	7.5
Gly	1.0	1	2.0	2	2.0	2	3.0	3	4.0	4	4.0	4	4.0	4.0	4.0	4.0
Asp	–	–	2.7	3	4.3	5	6.5	7	6.9	7	8.1	8	8.7	7.6	7.7	8.1
Ile	–	–	0.8	1	1.4	2	3.0	4	3.5	4	3.5	5	4.6	4.3	4.3	3.5
Thr	–		–		0.8	1	1.0	1	1.0	1	1.0	1	1.1	1.1	1.2	1.0
Phe	–		–		0.6	1	0.6	1	1.0	1	0.7	1	1.1	1.0	0.6	0.8
Ser	–		–		–		1.3	2	2.3	4	2.5	5	2.6	2.6	3.4	4.5
Tyr	–		–		–		–		–		0.1	1	0.1	0.2	–	–
Lys	–		–		–		–		–		0.2	1	0.2	n.b.	n.b.	n.b.
NH₃	–		–		–		–		7.4	5	7.6	5	5.4	n.b.	n.b.	n.b.
Cys(Bzl)	–		–		–		–		5.8	7	5.0	8	5.6	–	–	–
Cys(SO₃H)	–		–		–		–		–		–		–	0.7	8.2	7.2
Cys(SH)	–		–		–		–		–		–		–	4.0	–	–

* E = found experimentally on the product attached to the resin.

** T = theoretical.

I. Analysis of the S-benzyl-polypeptides following debenzylation in HBr-trifluoroacetic acid and subsequent hydrolysis in 6 N HCl, at 110°C, 0.2 torr, 30 hr.

II. Analysis of the synthetic S-sulfonate (hydrolysis conditions same as in I).

III. In order to obtain total cysteine, the S-sulfonate was oxidized with performic acid [12] prior to hydrolysis. Following this procedure, however, Tyr and Phe could no longer be quantitatively determined.

IV. Analysis of natural apoferredoxin. Oxidation with performic acid prior to hydrolysis (hydrolysis conditions same as in I. Time = 17 hr).

fonates from the natural and the synthetic apoferredoxin are not homogeneous, and the synthetic product shows even a more uniform chromatogram than the natural product. None of the isolated fractions F_1, F_2 or F_3 of natural apoferredoxin or synthetic apoferredoxin-sulfonate, being almost exclusively part of fraction F_2 could be reconstituted to active ferredoxin. However a combination of fractions F_1, F_2 and F_3 of natural apoferredoxin-

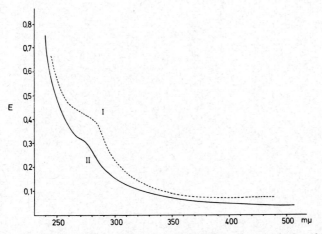

Fig. 4.   UV spectra of natural (I) and synthetic (II) apoferredoxin-S-sulfonate.

sulfonate can be reconstituted to active ferredoxin and it seems that fraction F_2 of the synthetic product can replace fraction F_2 of the natural product in these reconstitution-experiments. These reconstitution-experiments and the properties of natural and synthetic S-sulfonates are under further investigation.

The support of the Deutsche Forschungsgemeinschaft and the Welch Foundation in this work is gratefully acknowledged.

References

[1] M. Tanaka, T. Nakashima, A. Benson, H. Mower and K. T. Yasunobu, Biochemistry 5 (1966) 1666.
[2] E. Bayer, D. Josef, P. Krauß, H. Hagenmaier, A. Röder and A. Trebst, Biochim. Biophys. Acta 143 (1967) 435.
[3] E. Bayer et al., European J. Biochem.
[4] R. B. Merrifield, J. Am. Chem. Soc. 85 (1963) 2149; ibid. 86 (1964) 304; Biochemistry 3 (1964) 1385; J. Org. Chem. 29 (1966) 3100.

  [5] A. Marglin and R. B. Merrifield, J. Am. Chem. Soc. *88* (1966) 5051.
  [6] W. König, Dissertation, Universität Tübingen, unpublished.
  [7] R. Schwyzer, P. Sieber and H. Kappeler, Helv. Chim. Acta *42* (1959) 2622.
  [8] E. Schnabel, Liebigs Ann. Chem. *702* (1967) 188.
  [9] G. Jung, Dissertation, Universität Tübingen, July 1967.
 [10] Abkürzungen nach IUPAC, European J. Biochem. *1* (1967) 375.
 [11] H. Zahn, private communication.
 [12] J. L. Bailey, Techniques in protein chemistry (Elsevier, Amsterdam, 1967).
 [13] G. Pataki, Dünnschichtchromatographie (Walter de Gruyter and Co., Berlin).

ENZYMATIC SYNTHESIS OF PHAGE ΦX174 DNA

MEHRAN GOULIAN*

*The Pritzker School of Medicine of the University of Chicago, Departments of Medicine and Biochemistry, and the Argonne Cancer Research Hospital, U.S.A.***

For several years purified enzymes which can be used to synthesize DNA have been available from several sources, bacterial, viral and animal [1]. These enzymes, when mixed with the four deoxynucleoside triphosphates that correspond to the four bases in DNA, plus a DNA "primer", give rise to a product that resembles the DNA "primer" in a number of physical and chemical characteristics, including viscosity, solubility, optical properties, and susceptibility to deoxyribonucleases. The requirement for all four deoxynucleoside triphosphates, and the identity of product and primer in base ratio and nearest neighbour frequency, indicate that the primer DNA also serves as "template" in selection of bases in the product. It is natural to expect that if the synthetic product is indeed an accurate copy of the template DNA, biologic activity possessed by the template will also be conferred upon the synthetic copy. This expectation has been tested several times in the past using transformation systems [2–4]. In these studies, DNA with transforming activity was used as the template with purified DNA polymerase and deoxynucleoside triphosphates, and the synthetic product was examined for transforming activity. One of the problems that was encountered was the difficulty in effecting complete separation of the product from the template. This caused uncertainty about the source for any observed activity. A second problem was the dependence of specific transforming activity upon the size of the transforming DNA. Since there was no direct correspondence between the size of the synthetic DNA and the template with which it was to be compared, attempts to quantify results were greatly complicated. The results of these studies were inconclusive; they produced no certain evidence that there was product-associated transforming activity that could not be accounted for by contamination with template DNA.

* Most of the studies described in this paper were carried out while the author was working in the laboratory of Dr. Arthur Kornberg, Department of Biochemistry, Stanford University School of Medicine, with support from a National Institutes of Health Special Fellowship. Studies at the University of Chicago were supported in part by the Otho S.A. Sprague Memorial Institute, Chicago.
** Operated by the University of Chicago for the United States Atomic Energy Commission.

The reason for this apparent inability to synthesize transforming activity is not known, but it has been suggested that it is related to two anomalous features of synthetic DNA [3, 5]. One of these is the branched appearance that it presents in electron micrographs, which is quite unlike the simple linear form observed with natural DNA. The second is the tendency of synthetic strands to renature spontaneously upon removal of denaturing conditions, in contrast to natural DNA, which tends to remain denatured except under certain specific conditions. It has been suggested that the terminus of each branch in synthetic DNA consists of a loop joining the two strands. Such a structure would tend to renature spontaneously because of the extensive regions of self-complementarity within the molecule. This discussion will not examine possible reasons for the strange in vitro behavior of DNA polymerase which produces DNA with branches terminated by loops; suffice it to say that it is suspected that these properties were at the core of the transformation problem.

Our recent approach to the synthesis of biologically active DNA [6,7] utilized some of the properties of the bacteriophage ΦX174. The latter is a very small virus that infects the common colon bacillus *Escherichia coli*. The DNA of ΦX174 has a molecular weight of 1.6×10^6 dalton, corresponding to 5500 nucleotides which are in the form of a single-stranded circle. The phage particle consists of a single DNA molecule tightly packed inside a protein covering. Initial interest in ΦX174 for this kind of experiment arose from three considerations, one of which was the fact that there exists a system for demonstrating the biologic activity of ΦX174 DNA. Guthrie and Sinsheimer [8] have shown that the naked DNA isolated from ΦX174 can infect suitably modified cells of *E. coli*. When the cell wall of *E. coli* is made defective by special treatment (spheroplasting) this DNA is able to enter the cell, unaided by the viral protein coat. Penetration of the cell is followed by the production of many copies of complete phage particles within the bacterial spheroplast, as in normal infection by a whole phage particle. Infection occurs only with circular molecules of ΦX174 DNA; a single break, converting it to a linear structure, destroys infectivity. ΦX174 DNA met a second requirement with its ability to serve as template for purified DNA polymerase. Mitra et al. [9] had shown that DNA polymerase is able to use the single-stranded circular DNA as template, converting it to a duplex circle. The third aspect of the ΦX174 system which seemed to recommend it was the observation that the synthetic complementary ($-$) strand, produced on the circular viral ($+$) template, was not branched unless the enzyme was permitted to continue its action after converting the template to duplex [9].

Along with these favorable features there existed two formidable problems with the ΦX174 system: (1) it could not be expected that the synthetic (−) strand would be infective since it was linear rather than circular, and (2) even if it were circular there was no assurance that the (−) ring, a *complement* of the viral DNA, also was infective. The second objection was answered when it was learned both from the work of Rüst and Sinsheimer [10] and of Siegel and Hayashi [11], that (−) rings were infective. They had isolated (−) rings from the duplex ring structure (Replicative Form or RF) that occurs in the infected bacterial cells as an intermediate in the synthesis of viral DNA. The other objection was met by the discovery of enzymes, called polynucleotide joining enzymes, which link together suitably approximated ends of DNA. The existence of these enzymes was reported almost simultaneously by five different laboratories [12–16]. With the solution, by others, of these major problems, it seemed feasible to proceed to experimental testing of the ΦX174 system for synthesis of biologically active DNA. To recapitulate, the plan foresaw two stages, the first of which was the synthesis of duplex closed ΦX174 rings (RF) using viral (+) single-strand circular DNA as template and, second, the isolation from the RF of the synthetic (−) rings and assay of these for infectivity.

Synthesis of RF was approached in several ways, two of which were successful [6]. The first of these, although not used in subsequent studies, merits description because of the control it permitted at each step. This approach was motivated by the fear that polymerase action would not stop at one full replication of the ring template but would continue on and produce a branched unsuitable structure. Earlier data had suggested that bacteriophage T4 DNA polymerase, unlike the *E. coli* enzyme, could be expected to stop at one replication [17]. The problem with T4 polymerase was that, again unlike the *E. coli* polymerase, it could not initiate synthesis on the single-stranded circular template. From previous experience with the T4 enzyme, it could be expected to use the circular DNA as template if given a start, for example, by a short segment of (−) ring annealed to the (+) ring. A structure of this kind was achieved in the following way: *E. coli* polymerase was used with ΦX174 DNA templates to synthesize (−) strand material. The reaction was stopped after a relatively short time and before the amount synthesized was equivalent to the input template. The synthetic (−) strands, of varying length, were separated from the template with alkali and fractionated by sedimentation in an alkaline sucrose gradient. Fragments approximately one-fourth to one-eight the length of ΦX174 were selected and incubated with ΦX174 rings under conditions that would favor renaturation. Annealing was essentially complete, as judged by insusceptibility of the fragments to *E. coli* exonuclease I, which is specific for single-stranded DNA [18]. The resulting structure was the desired one, and when incubated with the T4 polymerase, it did serve as template, the 3′ OH ends of the small (−) fragments presumably functioning as sites for initiation of the polymerase. The T4 polymerase completed the (−) ring and stopped at one replication, as hoped. In fact, due to contamination with a minute amount of T4 polynucleotide joining enzyme, the (−) ring was closed to give the duplex closed ring structure, without requiring the addition of joining enzyme. Approximately 25–50% of the starting (+) rings bearing (−) ring fragments were converted to RF. This was determined in the manner discussed below and will not be described here.

The second method of synthesizing RF was much simpler although, in some of its details, it was not as well understood as the method just described. The success of the previous method encouraged the direct use of the *E. coli* polymerase for both initiation and completion of the synthesis. It was hoped that if the joining enzyme were present at the time of completion of the (−) ring, the latter would be closed and further synthesis of branched forms would thereby be prevented. The procedure consisted of incubating together ΦX174 DNA labeled with ^{3}H; *E. coli* DNA polymerase; the 4 deoxynucleoside triphosphates (one of them ^{32}P-labeled); *E. coli* joining enzyme; and DPN, the coenzyme for the joining enzyme. A small amount of the supernatant from a boiled crude extract of *E. coli* was also included in the mixture. When this was incubated for 3 hr at 20 °C, 30–50% of the (+) template rings were converted to RF. Evidence for such conversion was derived from three kinds of experiments. One of these employed centrifugation of the mixture with CsCl in the presence of the dye, ethidium bromide, using the procedure developed by Vinograd and coworkers [19]. Ethidium bromide intercalates between the bases of double-helical DNA with two effects that are pertinent here: the DNA becomes less dense, and the helix unwinds. Because it has no ends, circular DNA cannot unwind as much as linear DNA and, therefore, binds less dye and remains more dense. In a CsCl density gradient containing this dye, circular DNA molecules separate as a denser band from linear or nicked circular molecules that are otherwise of the same composition. The incubation mixture gave the two expected bands and fig. 1 shows that the heavier band, which would contain the RF, had an equivalent amount of template ^{3}H and product ^{32}P, within the accuracy of the experimental methods.

The structure was further confirmed by alkaline sucrose gradient sedimen-

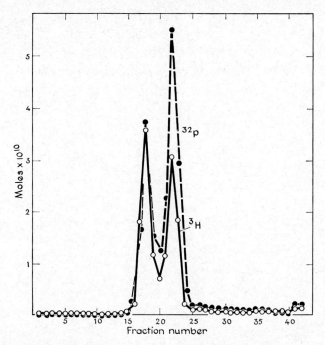

Fig. 1. Fractionation of duplex closed circles of ΦX174 DNA by CsCl density gradient centrifugation in the presence of ethidium bromide. The (+) circles were ^{3}H-labeled ΦX174 phage DNA. The ^{32}P-labeled synthetic (−) circles were synthesized with *E. coli* DNA polymerase and closed with *E. coli* polynucleotide joining enzyme. The peak on the left, which is at a higher density than the other peak, is made up of circular molecules.

tation. The two complementary circles in the RF cannot be separated, and in alkali they assume a complex and compact structure which has a distinctively high sedimentation rate. The DNA in the heavy band from CsCl-ethidium bromide centrifugation exhibited the expected rapid sedimentation in alkali and could not be distinguished from natural RF by this technique (fig. 2). The third technique used for characterization was electron microscopy. Electron micrographs showed simple unbranched circles with the expected contour lengths of 2 μ (in fig. 3).

All of the components of the incubation mixture were necessary for optimal yields of RF (table 1). There was an absolute requirement for the polymerase and triphosphates. Omission of DPN reduced the yield only slightly, presumably because the boiled extract also contained DPN. Significant amounts of RF were produced even in the absence of the joining enzyme, apparently because the polymerase preparation was contaminated with it. It

238

MEHRAN GOULIAN

TABLE 1

	Conversion of phage DNA to RF (%)
Complete system	38
Minus polymerase	< 0.2
Minus dTTP	< 0.2
Minus phage DNA	< 0.2
Minus DPN	26
Minus joining enzyme	4.5
Minus boiled extract	1.9

The complete system included phage DNA as template, *E. coli* DNA polymerase, the 4 deoxynucleoside triphosphates, *E. coli* polynucleotide joining enzyme and its coenzyme DPN, and a small amount of the supernatant from a boiled crude extract of *E. coli*. Conversion of the phage DNA to a closed duplex molecule was determined by alkaline sucrose gradient sedimentation as in fig. 2.

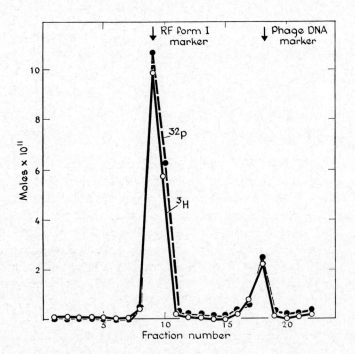

Fig. 2. Alkaline sucrose gradient sedimentation of purified partially synthetic duplex circular molecules. Material from the denser peak in fig. 1 was centrifuged in an alkaline sucrose gradient. Direction of sedimentation is to the left, and the position of natural duplex closed circles (RF-form I) is indicated at the top.

is interesting to note that the requirement for the boiled extract was even more stringent than the requirement for the joining enzyme. Although the exact function of the boiled extract was not known at the time, recent studies have provided some clarification. These studies indicate that the active material in the boiled extract consists of oligodeoxynucleotides derived from the *E. coli* DNA and that a small amount is actually incorporated into the synthetic DNA [21]. The oligodeoxynucleotides probably anneal to the template by fortuitous base pair homology and provide, at their 3' OH end, a site suitable for initiation of polymerase action [22]. The resemblance of this mechanism to the first method for synthesis of RF, described above, is self-evident.

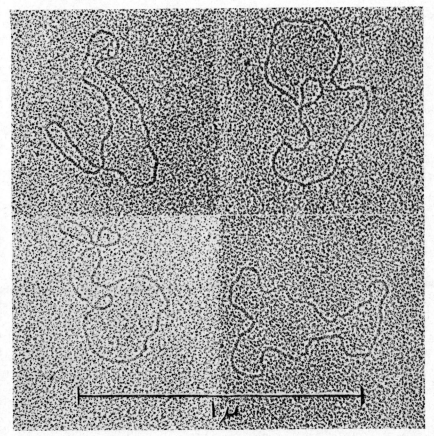

Fig. 3.   Electron micrographs of partially synthetic duplex closed circles. The source of the molecules was the denser peak in fig. 1. Preparation was based on the method of Kleinschmidt et al. [20].

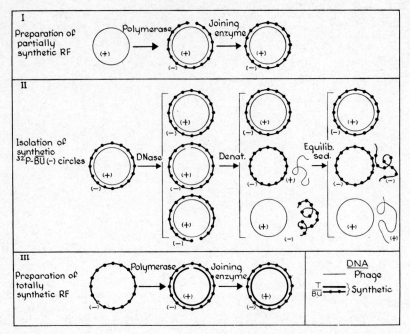

Fig. 4. Schematic representation of the preparation of synthetic (−) circles and RF.

Thus, with synthesis of RF, the first part of the plan was completed. It remained to isolate the (−) rings from the RF and test these for infectivity. For this purpose, RF was synthesized in the way described except that thymine was replaced with its analogue, bromouracil (BU) in the triphosphate mixture [6, 7]. DNA containing BU has a higher density than natural DNA, thereby facilitating the subsequent separation of product from template.

Since the two rings are interlocked, they cannot be separated without introducing at least one break into one of the rings. It would be most desirable to produce breaks in the (+) rings and thereby release intact (−) rings. Since no selective method of breakage is available, separation of the rings must be accomplished by introducing random nicks in both rings with, on the average, less than one nick per RF molecule. A similar technique had been used before, and the conditions chosen here employed pancreatic deoxyribonuclease in amounts sufficient to produce one break in half of the RF molecules (fig. 4). In this way, one-fourth of the molecules would be broken in the (+) ring leaving an intact (−) ring; one-fourth of them would be broken in the (−) ring, leaving an intact (+) ring; and one-half of them would be untouched. After denaturation, the mixture was centrifuged in a

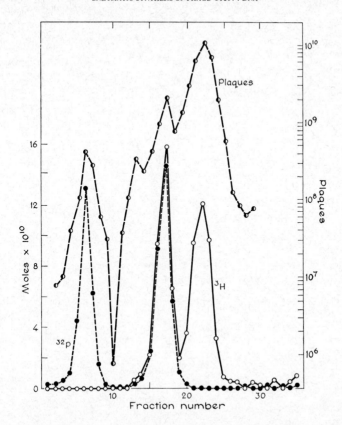

Fig. 5. Isolation of synthetic (−) ΦX174 DNA by density gradient centrifugation. Partially synthetic ΦX174 DNA in the form of duplex closed circles was prepared as shown in fig. 1 except that thymine was replaced by BU in the synthetic strand. This was treated with an amount of pancreatic DNase sufficient to cause one single-strand break in half the molecules, denatured with heat, and centrifuged to equilibrium in a CsCl density gradient. The most dense band, containing the synthetic ^{32}P DNA is on the left, the light ^{3}H-template DNA is on the right, and the intact duplex circles in the hybrid position, between the other two. The results of assays for infectivity by the spheroplast assay [8] are shown at the top ("Plaques").

CsCl density gradient, in order to separate it into 3 bands as follows: a heavy synthetic band, with BU and ^{32}P; the light, natural template material, with ^{3}H; and a hybrid band, with ^{32}P and ^{3}H, respresenting intact RF. This procedure gave the expected result, with three bands at the expected densities according to their composition (fig. 5). It was now of prime importance to determine whether or not the heavy band contained infectious material, and when the fractions were assayed for infectivity it was clear that the sought-for

result had been achieved: there was a distinct peak of infectivity corresponding to the heavy, synthetic band. The infectivity in the hybrid and light
bands was, of course, expected from their content of natural template.

To know that some synthetic ($-$) rings were infective was not enough; it
was necessary to know whether infection by the synthetic rings was comparable to infection by the natural material. If it were otherwise, it would have
been possible that the enzyme made many errors and that only a few molecules were correct copies of the template. This could not be determined
directly from the experiment just discussed since the content of broken rings
in the heavy and light peaks was unknown. Some of the reasons for this are
unreliability of pancreatic DNase in producing exclusively single-strand
scissions, the light sensitivity of BU-containing DNA, and chance contamination with nucleases. The circular and linear molecules were separated by
taking advantage of a slight difference in their sedimentation rates under
conditions of low salt. Using this method with material from both light and
heavy bands, it can be seen that, indeed, the predominant form is linear in
both but that the infectivity corresponds to the circular molecules evident in
each as a faster-moving shoulder (fig. 6). In addition, this peak of infective

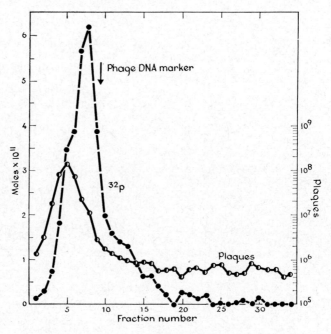

Fig. 6a

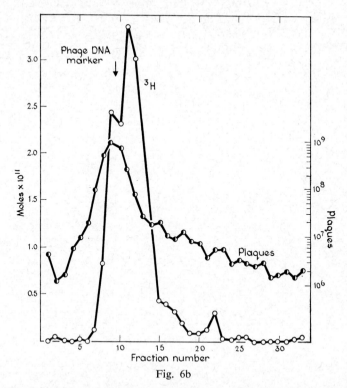

Fig. 6b

Fig. 6. Separation of synthetic and natural circular DNA by sedimentation. In (a) material from the heavy peak of the experiment shown in fig. 5 was centrifuged in a low salt sucrose gradient, (b) shows the same for the light peak (fig. 5). In each case, the faster moving shoulder represents the circular molecules, while the main peak contains broken (linear) forms.

circular molecules in the heavy band, due to its BU content, sedimented more rapidly than natural rings. From these data, it was now possible to calculate the specific infectivity of the synthetic molecules, i.e., the number of infected spheroplasts per mole of DNA, a number which compared favorably with that of their natural counterpart.

The conviction that the synthetic molecules were infective rests on 3 kinds of evidence: (1) the peak of infectivity in the CsCl density gradient (fig. 5), at the correct density for BU (−) rings, coincident with the ^{32}P of the synthetic material, clearly separated from adjacent peaks of infectivity, and lacking ^{3}H from template molecules; (2) physical separation of the synthetic molecules from template by a second method, in the low salt sedimentation experiment (fig. 6); and (3) distinctive light sensitivity of the synthetic

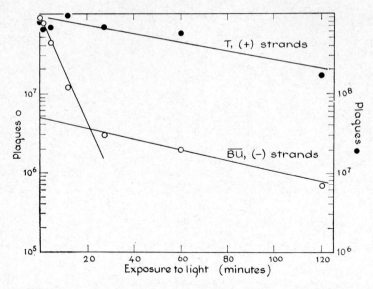

Fig. 7.   Inactivation by light of synthetic (BU) and natural DNA. Portions of the heavy
(^{32}P) and light (^{3}H) bands from the CsCl fractionation (fig. 5) were exposed to fluorescent
light and the infectivity followed with time of exposure.

molecules due to the BU (fig. 7). When samples from the heavy and light
CsCl bands were exposed to light under identical conditions the inactivation
of the heavy band was distinctly more rapid than that of the light band. This
is a property of BU-containing DNA and forces the conclusion that the infective molecules in that band contained BU and that they were, therefore,
synthetic.

It was possible to show that the synthetic ($-$) rings, like natural circular
DNA, could serve as templates for synthesis of RF. The same conditions
were used as described before except that the natural base thymine replaced
BU, and ^{3}H instead of ^{32}P was used to label the triphosphate. Again RF
was produced but now both rings were synthetic, the second generation
synthetic rings being ($+$), light, and labeled with ^{3}H. This fully synthetic
RF was infective (fig. 8) and, when it was treated with pancreatic DNase, as
described earlier, denatured, and sedimented in an alkaline sucrose gradient,
a peak of infectivity corresponding to the second generation synthetic rings
could be separated (fig. 9). These rings, rather than being complementary to
viral DNA, as in the previous experiment, now represented molecules that
were identical to the natural viral ($+$) DNA that had been used as template
for the first round of synthesis.

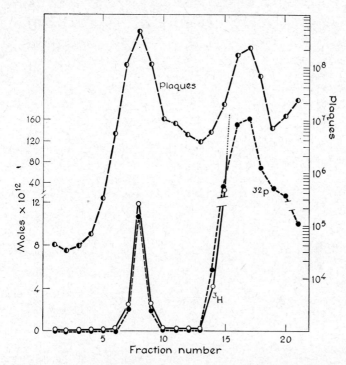

Fig. 8. Alkaline sucrose gradient sedimentation of fully synthetic RF. ^{32}P-BU synthetic rings (fig. 5) were used as template for synthesis of closed duplex circles as described in fig. 1, except that the triphosphates were labeled with ^{3}H. The mixture was sedimented in alkali and fractions were assayed for infectivity.

To summarize, purified DNA polymerase and polynucleotide joining enzymes from *E. coli* were used to convert the single-stranded circular DNA from phage ΦX174 to the duplex closed ring structure. The synthetic (−) rings in these molecules were isolated and found to be infective. The synthetic (−) rings served as template for a repeat of this procedure, yielding (+) rings which also were infective.

It should be pointed out that the level of performance demanded of the polymerase enzyme in these experiments was considerably greater than was the case in the original transformation experiments, which, it must still be assumed, failed for other reasons. The high level of precision of the polymerase considerably bolsters confidence in its ability to carry out gene replication in vivo. The system used here lends itself to some types of study not heretofore possible, such as, for example, the synthesis of biologically active DNA containing various base analogues, for the purpose of studying their mutagenic effects. It may be possible to study the mechanism for mutagenicity of certain mutant polymerases. It is also possible that this type of approach may prove useful in studies on mechanism of action of tumor viruses, especially on polyoma and SV 40, which in their physical features

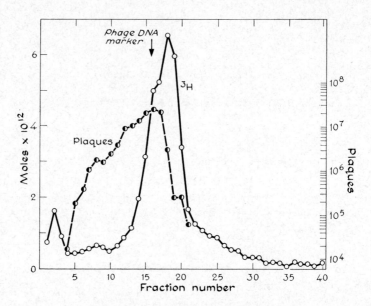

Fig. 9. Isolation of synthetic (+) rings by alkaline sucrose gradient centrifugation. Synthetic RF (fig. 8) was exposed to limited DNase action (fig. 4) and centrifuged in alkali. The faster moving shoulder, with corresponding peak of infectivity, corresponds to the synthetic (+) rings.

closely resemble ΦX174. The work has also encouraged hope that controlled alteration, or even addition of genes, may be possible on biologically active viral DNA for the ultimate purpose of introducing "corrections" into genetically defective human cells. These proposals remain purely speculative at present, but they do stimulate additional interest in amplifying this type of study and exploring details of mechanism, to permit additional understanding and control.

References

[1] A. Kornberg, Enzymatic synthesis of DNA, Ciba Lectures in Microbial Biochemistry (John Wiley & Sons, New York, 1961).
[2] R. M. Litman and W. Szybalski, Biochem. Biophys. Res. Commun. *10* (1963) 473.

[3] C. C. Richardson, C. L. Schildkraut, H. V. Aposhian, A. Kornberg, W. Bodmer and J. Lederberg, in Informational macromolecules, H. J. Vogel, V. Bryson and J. O. Lampen, eds. (Academic Press, New York, 1963) p. 13.

[4] C. C. Richardson, R. B. Inman and A. Kornberg, J. Mol. Biol. 9 (1964) 46.

[5] C. L. Schildkraut, C. C. Richardson and A. Kornberg, J. Mol. Biol. 9 (1964) 24.

[6] M. Goulian and A. Kornberg, Proc. Natl. Acad. Sci. U.S. 58 (1967) 1723.

[7] M. Goulian, A. Kornberg and R. L. Sinsheimer, Proc. Natl. Acad. Sci. U. S. 58 (1967) 2321.

[8] G. D. Guthrie and R. L. Sinsheimer, Biochim. Biophys. Acta 72 (1963) 290.

[9] S. Mitra, P. Reichard, R. B. Inman and A. Kornberg, J. Mol. Biol. 5 (1967) 424.

[10] P. Rüst and R. L. Sinsheimer, J. Mol. Biol. 23 (1967) 545.

[11] J. E. D. Siegel and M. Hayashi, J. Mol. Biol. 27 (1967) 443.

[12] M. Gellert, Proc. Natl. Acad. Sci. U.S. 57 (1967) 148.

[13] B. Weiss and C. C. Richardson, Proc. Natl. Acad. Sci. U.S. 57 (1967) 1021.

[14] B. M. Olivera and I. R. Lehman, Proc. Natl. Acad. Sci. U.S. 57 (1967) 1426.

[15] M. L. Gefter, A. Becker and J. Hurwitz, Proc. Natl. Acad, Sci. U. S. 58 (1967) 240.

[16] N. R. Cozzarelli, N. M. Melechen, T. M. Jovin and A. Kornberg, Biochem. Biophys. Res. Commun. 28 (1967) 578.

[17] M. Goulian, Z. J. Lucas and A. Kornberg, J. Biol. Chem. 243 (1968) 627.

[18] I. R. Lehman, J. Biol. Chem. 235 (1960) 1479.

[19] R. Radloff, W. Bauer and J. Vinograd, Proc. Natl. Acad. Sci. U.S. 57 (1967) 1514.

[20] A. K. Kleinschmidt, D. Lang, D. Jacherts and R. K. Zahn, Biochim. Biophys. Acta 61 (1962) 857.

[21] M. Goulian, Proc. Natl. Acad. Sci. U.S. 61 (1968) 284.

[22] M. Goulian, Cold Spring Harbor Symp. Quant. Biol. 33 (1968) 11.

REPEATED SEQUENCES IN DNA*

R. J. BRITTEN and D. E. KOHNE

*Department of Terrestrial Magnetism of the Carnegie Institution of Washington,
Washington D.C., U.S.A.*

The complementary structure of DNA plays a fundamental role in the cell. The complementary relations between nucleotide pairs are important not only in the duplication of DNA, but in the transcription and translation of genetic information. Matching of complementary nucleotide sequences is probably involved in genetic recombination as well as in other events of recognition and control within the cell.

It is a remarkable fact that separated complementary strands of *purified* DNA recognize each other. Under appropriate conditions they specifically reassociate [1]. This phenomenon has supplied a useful tool for exploring the nature of molecular events within the cell and broader biological questions such as the relationships among species [1–3].

Simple complementary ribopolymer pairs were shown in 1957 to form a helical paired structure when mixed in solution [4]. In 1960, DNA was dissociated into two strands, and the physical properties and biological activity of double-stranded DNA were then restored by incubation under appropriate conditions [1, 5]. In 1961, virus-specific RNA, made by bacteria during viral infection, was shown to pair with the viral DNA [6]. Techniques were developed for the immobilization of single-stranded DNA in cellulose [7], in agar [8], and on nitrocellulose filters [9]. It then became possible to assay the reassociation of radioactively labeled single-stranded fragments of DNA or RNA with the immobilized DNA.

Reassociation of the DNA of vertebrates was observed in 1964 [3]. The extent of reassociation between DNA strands derived from different species was shown to be a measure of the evolutionary relation between the species [10]. However, measurements also showed that the nucleotide sequence pairing was imprecise even when DNA from a single species was reassociated [11].

Before these measurements were made it had been expected that it would be very difficult to observe the reassociation of the DNA of vertebrates and other higher organisms [1]. The enormous dilution of individual nucleotide sequences in the large quantity of DNA in each cell was expected to make

* Reprinted by permission from Science *161* (1968) 529.

the reaction so slow that months would be required for its completion at practical concentrations with the DNA-agar method.

Investigation of this paradox was begun in our laboratory in early 1964, and shortly afterward the hypothesis was put forward [12] that some nucleotide sequences were frequently repeated in the DNA of vertebrates. This supposition was supported by the observation that 10 percent of the DNA of the mouse reassociated extremely rapidly. This fraction, identified as mouse-satellite DNA was shown by later measurements [13] to consist of a million copies of a short nucleotide sequence [13]. Later work [14, 15] has shown that repeated nucleotide sequences are of very general occurrence.

In this article we describe selected measurements [12–15] that show most clearly the presence of repeated sequences and indicate some of their properties.

Conditions for reassociation

The conditions for efficient reassociation of DNA were explored originally by Marmur et al. [1] and have since been studied in several laboratories [15–17]. Briefly stated, the requirements are as follows. (i) There must be an adequate concentration of cations. Below 0.01 M sodium ion, the reassociation reaction is effectively blocked. (ii) The temperature of incubation must be high enough to weaken intrastrand secondary structure. The optimum temperature for reassociation is about 25 °C below the temperature required for dissociation of the resulting double strands. (iii) The incubation time and the DNA concentration must be sufficient to permit an adequate number of collisions so that the DNA can reassociate. (iv) The size of the DNA fragments also affects the rate of reassociation and is conveniently controlled if the DNA is "sheared" to small fragments *. Thus, in order to achieve reproducible reassociation reactions the cation concentration, temperature of incubation, DNA concentration, and DNA fragment size must all be controlled **.

* The DNA used in this work has been sheared to a relatively uniform population of small fragments (about 400 or 500 nucleotides long) by passing it twice through a needle valve with a pressure drop of 3.4 kilobar.

A specially built air-operated plunger pump was used to develop this high pressure. The DNA is denatured when sheared at 3.4 kilobar, unless a very high salt concentration is present to raise the temperature of melting (T_m). These small fragments give reproducible rates of reassociation and do not, under the usual condition for reassociation, form large aggregates or networks.

** For some purposes, correction can be made for different values of the parameters. There have been some quantitative measurements of the effect of temperature [1, 15] salt concentration [15–17] and fragment size [16, 17].

The measurement of reassociation

Reassociation can be measured in a variety of ways, each depending on some easily detected physical difference between single-stranded (dissociated) DNA and double-stranded (reassociated) DNA [1]. For example, dissociated DNA absorbs more ultraviolet light than reassociated DNA does. Double-strand DNA also has a greater degree of optical activity than single-strand DNA.

In the DNA-agar method, reassociation is monitored by measuring the binding of labeled fragments of single-stranded DNA to long strands of DNA physically immobilized in a supporting substance. The immobilization prevents reassociation of the long DNA with itself. After incubation the unbound fragments are washed away, and the quantity of bound radioactive fragments is measured. It is now possible to measure the reassociation of DNA fragments with DNA immobilized on nitrocellulose filters [18]. The rate of the reaction is markedly reduced compared to the rate in solution [19].

Another useful technique for measuring reassociation depends on the fact that double-stranded DNA can be separated from single-stranded DNA on a calcium phosphate (hydroxyapatite) column [20]. Reassociation reactions can be followed by passing samples through hydroxyapatite and determining the amount of double-stranded DNA adhering to the column. This technique is particularly useful since DNA can be fractionated on a preparative scale on the basis of its ability to reassociate at a given C_0t, a parameter which may be explained as follows.

The meaning of C_0t

Much of the evidence for repeated sequences depends on measurements of the rate of reassociation. In addition, the design of most reassociation experiments is strongly influenced by the time required to complete the process. The reassociation of a pair of complementary sequences results from their collision, and therefore the rate depends on their concentration. The product of the DNA concentration and the time of incubation is the controlling parameter for estimating the completion of a reaction. For convenience and simplification of language we have chosen to call this useful parameter C_0t, which is expressed in moles of nucleotides times seconds per liter*.

---

* Here, C_0t is used as a noun and may be pronounced as the homonym of "cot." A C_0t of 1 mole × sec/liter) results if DNA is incubated for 1 hr at a concentration of 83 μg/ml, which corresponds to an optical density of about 2.0 at 260 nm.

Evidence is presented below that the DNA of each organism may be characterized by the value of C_0t at which the reassociation reaction is half completed under controlled conditions. The rates observed range over at least eight orders of magnitude. Therefore we have found it necessary to introduce a simple logarithmic method for the presentation of measurements of reactions over extended periods of time and wide ranges of concentration. For illustration, fig. 1 shows the progress of an ideal second-order reaction

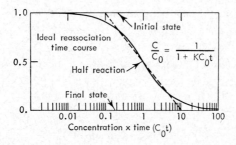

Fig. 1. Time course of an ideal, second-order reaction to illustrate the features of the $\log C_0t$ plot. The equation represents the fraction of DNA which remains single-stranded at any time after the initiation of the reaction. For this example, K is taken to be 1.0, and the fraction remaining single-stranded is plotted against the product of total concentration and time on a logarithmic scale.

plotted as a function of the product of the time of reaction and the DNA concentration on a logarithmic scale. On such a graph, reactions carried out at different concentrations may be compared, and the data may be combined to give a complete view of the time course of the reaction.

The symmetrical shape of an ideal second-order curve plotted in this way is pleasing and convenient. The central two-thirds of the curve follows closely a straight line, shown dashed. One useful indicator is the slope of this line which can be evaluated from the ratio of the values of C_0t at its two ends. This ratio is about 100 for an ideal reaction when estimated as shown on fig. 1. If the ratio is much greater than 100, the reaction is surely heterogeneous; that is, species with widely different rates of reassociation are present.

Rate of reassociation of DNA

Reassociation of a pair of complementary strands results from their collision. Therefore we expect the half-period for reassociation to be inversely proportional to the DNA concentration under fixed conditions for a particular DNA *.

Further, one would expect for a given total DNA concentration that the half-period for reassociation would be proportional to the number of different types of fragments present and thus to the genome size **. This expectation is exactly borne out in several cases. In fig. 2, the time course of reassociation of a number of double-stranded nucleic acids is shown. Within

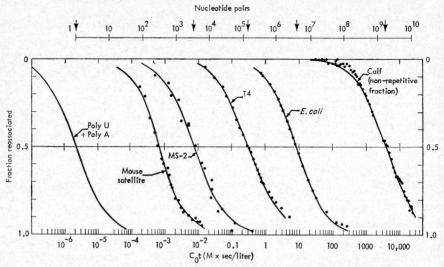

Fig. 2. Reassociation of double-stranded nucleic acids from various sources. The genome size is indicated by the arrows near the upper nomographic scale. Over a factor of 10^9, this value is proportional to the C_0t required for half reaction. The DNA was sheared and the other nucleic acids are reported to have approximately the same fragment size (about 400 nucleotides, single-stranded). Correction has been made to give the rate that would be observed at 0.18 M sodium-ion concentration. No correction for temperature has been applied as it was approximately optimum in all cases. Optical rotation was the measure of the reassociation of the calf thymus nonrepeated fraction (far right). The MS-2 RNA points were calculated from a series of measurements [23] of the increase in ribonuclease resistance. The curve (far left) for polyuridylic acid + polyadenylic acid was estimated from the data of Ross and Sturtevant [24]. The remainder of the curves were measured by hypochromicity at 260 nm; a Zeiss spectrophotometer with a continuous recording attachment was used.

* In our experience, the reassociation of purified sheared DNA shows the concentration dependence expected for a second-order reaction. For DNA without repeated sequences, the time course also approximately follows second-order kinetics. While earlier measurements have suggested greater complexity, this is not supported by more recent work [17].
** The word genome customarily means the genetic constitution of an organism. Here the genome size is taken to mean the haploid DNA content of a cell or virus particle. The number of different fragments will only be proportional to the genome size in the absence of repetition or unrecognized polyploidy.

the precision of the measurements, the reassociation of these various DNA's follows the time course of a single second-order reaction. In each case where it is applicable, the genome size is marked with an arrow on the upper scale. Cairn's measurement* [21] of the size of the *Escherichia coli* genome (4.5×10^6 nucleotide pairs) has been used to fix and locate this scale. The length of the T2 bacteriophage chromosome has also been carefully measured and found to be 2×10^5 nucleotide pairs, and the size of T4 is similar [22]. The total length of MS-2 viral RNA is 2.4×10^6 or 4000 nucleotide pairs in the double-stranded replicative form [23]. The rate of reassociation [24] of the homopolymer pair [polyuridylic acid plus polyadenylic acid (polyU + polyA)] is consistent with the fact that these molecules are complementary in all possible registrations.

The proportionality between the C_0t required for half-reassociation of the DNA and the genome size is only true in the absence of repeated DNA sequences.

Fig. 2 also shows the time course of reassociation for two fractions isolated from mammalian DNA. These fractions both follow the curve expected for a single second-order reaction, but one fraction reassociates more rapidly than the smallest virus, while the other reassociates 500 times more slowly than bacterial DNA. The former (mouse satellite DNA) represents 10 percent of the mouse DNA; its rate of reassociation indicates that the segment is roughly 300 nucleotide pairs and must be repeated about a million times [13] in a single cell. At the other extreme is a slowly reassociating fraction which includes about 60 percent of calf DNA. Its rate of reassociation is just that expected if it were made of unique (non-repeating) sequences. The calf genome contains 3.2×10^9 nucleotide pairs [25].

Repeated sequences in the DNA of calf and salmon

In order to obtain a fairly complete view of the repeated sequences in one organism, it is necessary to measure the degree of reassociation over a very wide range of C_0t. In fig. 3 the reassociation of calf thymus DNA measured by the hydroxyapatite procedure is shown. The hydroxyapatite method is convenient for this purpose since the degree of reassociation can be directly determined by assay of the amount of DNA which is bound [20, 15]. Sam-

* Because of its apparent lack of repetition and measured genome size *E. coli* DNA is used as a reference for the comparison of reassociation rates of other DNA. The use of this reference in future work will permit comparison of rates where fragment size and other conditions affecting absolute rate may vary.

R. J. BRITTEN AND D. E. KOHNE

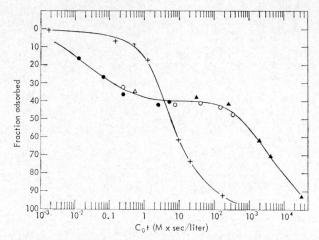

Fig. 3.   The kinetics of reassociation of calf-thymus DNA measured with hydroxyapatite. The DNA was sheared at 3.4 kilobars and incubated at 60°C in 0.12 M phosphate buffer. At various times, samples were diluted, if necessary (in 0.12 M phosphate buffer at 60°C), and passed over a hydroxyapatite column at 60°C. The DNA concentrations during the reaction were (μg/ml): open triangles, 2; closed circles, 10; open circles, 600; closed triangles, 8600. Crosses are radioactively labeled *E. coli* DNA at 43 μg/ml present in the reaction containing calf thymus DNA at 8600 μg/ml.

ples were simply diluted into a convenient volume of 0.12 M phosphate buffer* and passed over hydroxyapatite in a water-jacketed column at 60°C. A variety of tests [15, 26] have shown that under these conditions reassociated DNA is quantitatively bound, while not more than 0.5 percent of single-stranded DNA is adsorbed. The concentration of DNA present in the incubation mixture also can be varied over a wide range without interfering with the determination.

The hydroxyapatite binding measurements (fig. 3) show that 40 percent of the calf DNA has reassociated before a C_0t of 2. Little if any reaction occurs in the next two decades of C_0t. Thus for calf DNA, there is a clear separation between DNA which reassociates very rapidly and that which reassociates very slowly.

The rapidly reassociating fraction in calf DNA requires a C_0t of 0.03 for half-reassociation, whereas the slowly reassociating fraction requires a C_0t of 3000. Thus the concentration of DNA sequences which reassociate rapidly is 100,000 times the concentration of those sequences which reasso-

* This phosphate buffer is composed of an equimolar mixture of Na_2HPO_4 and NaH_2PO_4. The indicated molarity is for the phosphate. The sodium-ion concentration is 1.5 times greater.

ciate slowly. If the slow fraction is made up of unique sequences, each of which occurs only once in the calf genome, then the sequences of the rapid fraction must be repeated 100,000 times on the average.

The measurements shown on fig. 3 were done in several series at different DNA concentrations. Nevertheless, the results are concordant. The points fall on a single curve with good accuracy. This establishes that the measured reassociation process results from a bimolecular collision. In turn, the rapidity of the early part of the reassociation reaction can result only from high concentrations of the reacting species. We may conclude that about 40 percent of calf DNA consists of sequences which are repeated between 10,000 and a million times.

In one of the series of measurements shown on fig. 3, in addition to the 8600 μg of sheared calf DNA per milliliter there was present 43 micrograms of ^{32}P-labeled sheared E. coli DNA per milliliter, serving as an "internal standard". The simultaneous assay of the reassociation of the two DNA's that are present together controls a variety of possible experimental errors*.

Fig. 4 shows the reassociation of salmon sperm DNA measured with hydroxyapatite. Most of the salmon DNA appears to be made up of repeated sequences. The average degree of repetition is not as great as it is for the repeated fraction of calf DNA. The fact that the major part of the process extends over more than a factor of 10,000 in C_0t shows that many different degrees of repetition are present, varying from perhaps 100 copies to as many as 100,000 copies. The reassociation of the unique (single copy) DNA of this organism has not yet been observed. It would be expected to reassociate with a C_0t greater than 1000, and no measurements have yet been made in this region for salmon DNA.

The occurrence of repetitious DNA

With the observation of repeated DNA sequences in several vertebrate genomes the question arises: are the DNA's of these creatures exceptional,

---

* The variables controlled in this way are salt concentration, temperature, and viscosity. In addition, any possible nonspecific interactions in the DNA at this high concentration will have similar effects on both the E. coli and calf DNA reassociation reactions. The half reaction C_0t for E. coli DNA in fig. 3 is 6.0, whereas on fig. 2 it is 8.0. Reactions usually appear twofold faster when assayed with the hydroxyapatite method as compared to the optical method since the fraction of fragments reassociated is measured in one case, while the fraction of total strand length reassociated is measured in the other case [15]. On fig. 3 there may be a 50 percent increase in the C_0t for half reaction for the data taken at 8600 μg/ml. This decrease in the rate of reassociation is due to the increased viscosity of the incubation solution.

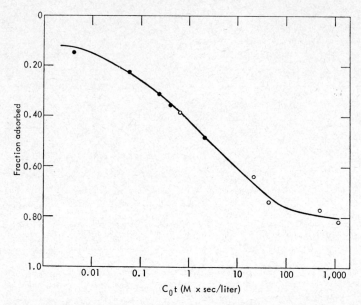

Fig. 4. The kinetics of reassociation of salmon sperm DNA measured with hydroxy-apatite. The DNA was sheared at 3.4 kilobar and incubated at 50 °C in 0.14 M phosphate buffer. The samples were diluted into 0.14 M phosphate buffer at 50 °C and passed over hydroxyapatite at 50 °C. The DNA concentrations during the incubation were (μg/ml): closed circles, 8; open circles, 1600.

or do repeated sequences occur generally among higher organisms? A limited survey was therefore carried out by the following procedure. (i) DNA was prepared and purified * from many organisms and then sheared to fragments consisting of 500 nucleotides. (ii) The DNA was dissociated in 0.12 M phosphate buffer and incubated at C_0t of 1 to 10 at 60 °C. (iii) The solution was passed over a hydroxyapatite column equilibrated to 0.12 M phosphate buffer and 60 °C [15, 26]. Under these conditions only the reassociated DNA becomes adsorbed to the column. (iv) The adsorbed DNA was eluted, and its reassociation kinetics were measured in the spectrophotometer.

The optical reassociation measurements for rapidly reassociating fractions prepared in this way from DNA of four organisms are shown in fig. 5. All

---

* The DNA was prepared by a combination of methods [27]. Purity was tested in the spectrophotometer by melting in 0.12 M phosphate buffer. We required that there be no measurable rise in optical density between 40° and 70 °C, and that a normal melting curve was obtained with a hyperchromicity of at least 25 percent of the absorbancy at 98 °C. Commercial calf and salmon DNA were utilized in some experiments, and no differences were observed with results obtained with DNA prepared from fresh tissue.

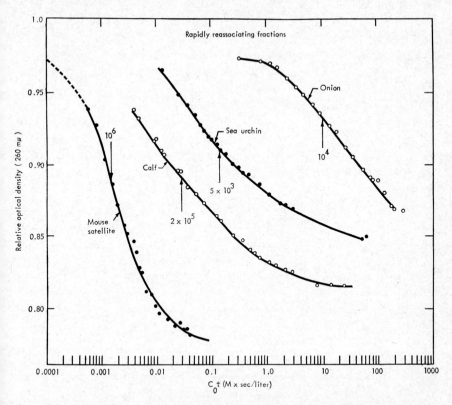

Fig. 5. Optical reassociation curves of repetitious DNA fractions from various organisms. All of the fractions were purified on hydroxyapatite with only minor modifications in the procedure for each different DNA. The left scale gives the ratio of the optical density at 60°C to the initial value measured at 98°C. All of the fractions except that from onion were reassociated in 0.08 M phosphate buffer at 60°C. The onion DNA was reassociated in 0.24 M phosphate buffer at 60°C. The onion points were plotted a factor of 5 to the right to allow for the increased rate of reassociation, and give approximately the curve that would be observed in 0.08 M phosphate buffer. The arrows permit estimation of the average degree of repetition in each case. They are located at the C_0t at which a fraction with the indicated degree of repetition would be half reassociated. The genome size and the amount of the rapidly reassociating fraction enter into the calculation in each case.

of these organisms contain repetitive DNA. The rate of reassociation of these fractions is very much faster than that calculated from the respective genome sizes. However, the reassociation pattern observed is quite different in each of the four cases. The curves for sea urchin DNA and calf DNA are probably representative of most of the repeated DNA in these organisms. However, the curve for mouse satellite DNA represents only the most repetitive fraction of the DNA from mouse cells [13], since the C_0t used before fractionation was very small. In the case of the onion DNA there is another repeating fraction which reacts more slowly than the fraction from onion shown in fig. 5. This more slowly reacting fraction appears to have a repetition frequency between 100 and 1000.

Results obtained from DNA-agar experiments [2, 3] expand the list of higher organisms which contain repetitious DNA. The reassociation conditions of the DNA-agar technique yield a C_0t between 1 and 100. For DNA from higher organisms only repeated sequences will reassociate appreciably at these C_0t's. Therefore, the reassociation detected in DNA of higher organisms by the DNA-agar technique has been due to the reassociation of repetitious DNA. A list of organisms in which repeated DNA sequences have been found is shown in table 1. Since so many types of organisms are represented it seems virtually certain that repetitious DNA is universally present in higher organisms. In assembling this table we have made use of several sets of results [3, 28].

The species of bacteria (*E. coli, Clostridium perfringens, Proteus mirabilis*) that have been examined do not contain repetitious DNA detectable by our methods. In none of these cases was reassociated DNA of low thermal stability observed. In all cases the kinetic curve for reassociation apparently contained only the one major component. As a further check for repetitious DNA in *E. coli*, the first small fraction to reassociate ($C_0t = 0.5$) was isolated on hydroxyapatite and shown to reassociate at the same rate as most of the *E. coli* DNA. While the sensitivity of the test is high, the existence of a small amount of repetitious DNA cannot be ruled out. Optical measurement of the reassociation kinetics on unfractionated DNA from several viruses (simian virus 40 and bacteriophages T4 and lambda) has likewise given no evidence of repeated DNA sequences.

Only a very small repetitive fraction has been detected in DNA from *Saccharomyces cerevisiae*. Because of its small quantity and low native thermal stability it can tentatively be identified as mitochondrial DNA. At this moment, the relatively fragmentary evidence suggests that eukaryotes (except possibly yeast) contain repetitious DNA while prokaryotes do not. A great number of measurements will be necessary to ascertain the boundary between those life forms which do and those which do not have repetitious DNA.

Table 1 also describes interactions of DNA from a variety of tissues. We have seen no evidence for a variation in the pattern or amount of repeated sequences between different tissues of a given species or individuals of a species. In this work, sensitive tests for differences were not made; however,

TABLE 1
Occurrence of repetitious DNA

Protozoa
  Dinoflagellate (*Gyrodinium cohnii*)*
  *Euglena gracilis**

Porifera
  Sponge (*Microciona*)*

Coelenterates
  Sea anemone (*Metridium*) (tentacles)*

Echinoderms
  Sea urchin (*Strongylocentrotus*)
  (sperm)*†‡
  Sea urchin (*Arbacia*) (sperm)*†‡
  Starfish (*Asterias*) (gonads)*
  Sand dollar (*Echinarachnis*)‡

Arthropods
  Crab (*Cancer borealis*) (gonads)*
  Horseshoe crab (*Limulus*)
  (hepatopancreas)*

Mollusks
  Squid (*Loligo pealii*) (sperm)*

Elasmobranchs
  Dogfish shark (liver)*

Osteichthyes
  Salmon (sperm)*†‡
  Lungfish*†

Amphibians
  Amphiuma (liver, red blood cells, muscle)*
  Frog (*Rana pipiens*)†
  Frog (*Rana sylvatica*)‡
  Toad (*Xenopus laevis*) (heart, liver, red blood cells)
  Axolotl (*Ambystoma tigrinum*)‡
  Salamander (*Triturus viridescens*)‡

Birds
  Chicken (liver, blood)*†‡

Mammals
  Tree shrew‡
  Armadillo‡
  Hedge hog‡
  Guinea pig‡
  Rabbit‡
  Rat (liver)*†‡
  Mouse (liver, brain, thymus, spleen, kidney)*†‡
  Hamster‡
  Calf (thymus, liver, kidney)*†‡

Primates
  Tarsier‡
  Slow Loris‡
  Potto‡
  Capuchin‡
  Galago‡
  Vervet‡
  Owl monkey‡
  Green monkey‡
  Gibbon‡
  Rhesus†‡
  Baboon‡
  Chimpanzee*‡
  Human*†‡

Plants
  Rye (*Secale*)‡
  Tobacco (*Nicotiana glauca*)‡
  Bean (*Phaseolus vulgaris*)‡
  Vetch (*Vicia cillosa*)‡
  Barley (*Hordeum vulgare*)*†
  Pea (*Pisum sativum* var. Alaska)*†
  Wheat (*Triticum aestivum*)*‡
  Onion (*Allium* sp.)*

* Rate of reassociation measured directly by hydroxyapatite fractionation or measurement of optical hypochromicity as a function of time or both.
† Labeled, sheared fragments bind to DNA from the same species embedded in agar at a C_0t so low that repetition must be present.
‡ Sheared nonradioactive fragments of DNA from the listed organism complete with the DNA-agar reaction (†) of a related species, reducing the amount of labeled DNA which binds to the embedded DNA.

earlier experiments of McCarthy and Hoyer [29] with the DNA-agar method were specifically designed to detect variation of DNA from tissue to tissue. These may now be interpreted as showing that the repeated sequences in the mouse DNA occur to about the same extent in many tissues and in cultured cell lines.

The precision of repetition

When DNA strands which are not perfectly complementary reassociate, the

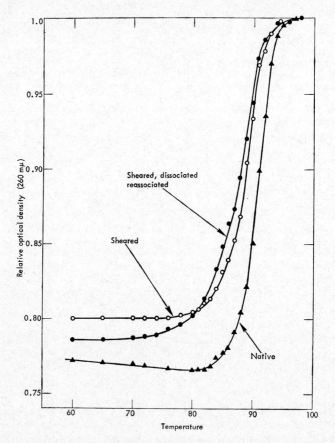

Fig. 6.   Melting curves of *E. coli* DNA in 0.12 M phosphate buffer. Open circles, native DNA sheared at 3.4 kilobar; closed circles, similarly sheared DNA dissociated (100°C, 5 min) and reassociated by incubation at 60°C in 0.12 M phosphate buffer; triangles, native unsheared DNA. Shearing at 3.4 kilobar dissociates a part of the DNA, accounting for the somewhat greater hyperchromicity of the reassociated DNA.

resulting pairs have reduced stability. This effect supplies a method for measuring the degree of sequence difference among DNA strands. Measurements with artificial polymers indicate that, when 1 percent of the base pairs are not complementary, the temperature at which dissociation occurs (melting temperature, T_m) is about 1 °C lower than that for perfectly complementary strands [30, 12]. The data of fig. 6 show that shearing bacterial DNA to small fragments, and its dissociation and reassociation, do not have a

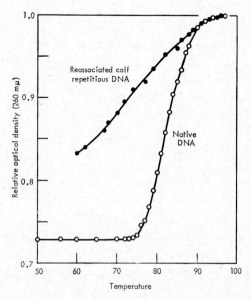

Fig. 7.   DNA melting curves in 0.08 M phosphate buffer. Open circles, unsheared native calf DNA; closed circles, reassociated calf repetitious DNA, sheared at 3.4 kilobar. In this solvent, single-stranded DNA gives an absorbancy change of only 3 percent from 60 to 95 °C (see fig. 9).

large effect on the melting temperature. This means that in the helically paired regions virtually all of the bases in reassociated bacterial DNA are properly matched. Thus the strong reduction in thermal stability observed for calf DNA indicates actual dissimilarities in the paired sequences of the reassociated DNA.

Fig. 7 shows the change in adsorbance with temperature of a repetitious fraction of calf thymus DNA (prepared by hydroxyapatite fractionation of DNA sheared at 3.4 kilobar). The optical density of this fraction changes over a wide range of temperature. Most of the thermal dissociation occurs below the temperature at which native DNA begins to dissociate. The change of absorbancy shows that base-paired structure has been formed in significant amount. The broad range of dissociation, in turn, indicates imprecise pairing. This observation confirms and extends earlier measurements with DNA-agar [11] in which more reassociation occurred at lower temperatures of incubation both for intraspecies and interspecies pairs. The temperature and salt concentration during a reassociation incubation establish a criterion of precision in that pairs form only if they are stable above the incubation temperature. Thus the incubation temperature determines which set of sequences will reassociate and controls the resulting melting temperature.

Raising the temperature of hydroxyapatite causes adsorbed double-stranded DNA to dissociate. The resulting single-stranded fragments may then be eluted. When dissociation is plotted against temperature, the profiles are very similar to those measured by change in ultraviolet absorbancy in free solution [15, 20].

Fig. 8 shows reassociated repetitive salmon DNA fractionated with hydroxyapatite on the basis of its thermal stability. Sheared DNA from salmon was dissociated and then incubated at 50 °C in 0.14 M phosphate buffer for

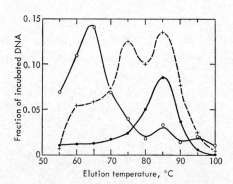

Fig. 8. Thermal fractionation on hydroxyapatite of reassociated salmon sperm DNA. DNA sheared at 3.4 kilobar was incubated at 50°C in 0.14 M phosphate buffer (C_0t, 370) and passed over hydroxyapatite at 50°C in 0.14 M phosphate buffer. The adsorbed DNA was eluted by exhaustive washing (0.14 M phosphate buffer) at intervals of 5 °C (dashed line and crosses). To show specificity, four fractions (65, 70, 85, 90°C) were again denatured (100°C, 5 min) and reincubated (50°C, 0.14 M phosphate buffer, C_0t about 10) and readsorbed on hydroxyapatite at 50°C. Two of these were again thermally eluted from a column: open circles, 65 °C fraction; closed circles, 85 °C fraction. The other two were eluted with 0.4 M phosphate buffer and melted in the spectrophotometer as shown in fig. 9.

a C_0t of 270, and the reassociated DNA was adsorbed on hydroxyapatite at 50 °C, 0.14 M phosphate buffer. The temperature of the column was raised in steps of 5 °C, and at each temperature the dissociated DNA was washed from the column with 0.12 M phosphate buffer. The resulting chromatogram (fig. 8, dashed line) shows a broad range of thermal stability. In order to establish the specificity of the fractionation, samples eluted at 65 and 85 °C were incubated again at 50 °C. They were then readsorbed and reanalyzed as before.

The strand pairs formed during the second incubation are ordinarily not the ones that were originally eluted. Instead, they are new duplexes formed by randomly assorted pairings among the selected set of strands. In each case, however, the same average degree of precision of relationship results. The portion eluting at 65 °C shows a peak again at 65 °C, and the 85 °C portion peaks at 85 °C. The degrees of sequence divergence are thus characteristic of these sets of fragments. Similar studies have been done with calf thymus DNA with entirely comparable results. In addition, experiments with labeled calf DNA fractions indicate that little sequence homology exists between precisely and imprecisely reassociating sets of repetitive DNA.

Length of repeated sequences

Are reassociated repeated sequences complementary only in short regions or are they complementary over most of their length? The thermal stability of a pair does not by itself answer this question since it appears that complementary sequences, 100 nucleotide pairs long, will have a thermal stability approaching that of very long complementary sequences [19, 31]. However, ultraviolet hyperchromicity is a measure of the extent of sequence matching. Therefore, the hyperchromicity of a preparation of strand pairs gives a measure of the fraction of the total length which is complementary. Results for two such preparations are shown in fig. 9. Native, completely complementary salmon DNA has a hyperchromicity of about 0.25. Single-stranded DNA has a hyperchromicity of 0.06 and melts mostly at lower temperatures, as shown by the top curve.

The 70 and 90 °C fractions each have about half the hyperchromicity of native DNA. From this we may conclude in each case that the base-paired regions of the reassociated repetitious DNA include about half of the nucleotides of the fragments.

Several complicating factors interfere with a more firm conclusion. Reassociated fragments will, in general, have single-stranded ends, since two

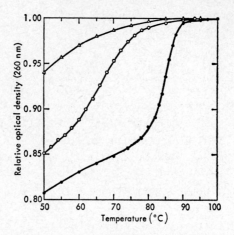

Fig. 9. Spectrophotometric melting curves in 0.14 M phosphate buffer of fractions of salmon sperm DNA. Fractions were prepared as described in fig. 8: closed circles, fraction eluted at 90°C; open circles, fraction eluted at 70°C. The upper curve (open triangles) is for the DNA which did not bind to hydroxyapatite (50°C, 0.14 M phosphate buffer) in the first incubation and is therefore purely single stranded.

complementary fragments rarely terminate at the same points in the sequence. All degrees of overlap will occur, and for first pairing the expected hyperchromicity is between one-half and two-thirds that for native DNA. We do not know the further extent of pair formation involving the single-stranded ends in these preparations. Finally, of course, repeated sequences have diverged from each other, and the unmatched nucleotides occurring within the paired sequences reduce the hyperchromicity.

These measurements are corroborated by the hyperchromicity we have observed for reassociated repetitive DNA fractions from many organisms. It usually falls between 0.10 and 0.20. Some examples are shown on fig. 5. The few CsCl equilibrium centrifugation measurements we have made show a marked decrease in density upon reassociation which also implies a good extent of complementary pairing of repetitive DNA.

It appears that, on the average, repeated sequences are not extremely short (not less than 200 nucleotides) and may be much longer than our fragments, which average perhaps 400 nucleotides. In other words, wherever a region of sequence homology occurs between two fragments of the genome, it is likely to continue for at least several hundred nucleotides. It does not usually continue perfectly, however, since the reduced thermal stability observed implies that local interruptions of the homology must be scattered through the regions of homology.

Nonrepeated DNA of higher organisms

Somewhat more than half of the DNA from mouse or calf may be recovered in the single-stranded state after an extensive incubation (fig. 3; $C_0t = 100$). The subsequent, very slow, reassociation of this fraction has been measured by three methods – hydroxyapatite adsorption (fig. 3), optical rotation increase (370 nm) (fig. 10), and hypochromicity (260 nanometers). The measurements establish that this fraction reassociates accurately and nearly completely. Experiments [15] indicate species dependence and therefore sequence specificity of the reassociation of the very slow fraction.

The curves shown on fig. 3 give the results of a measurement of the rate of reassociation of the slow fraction of calf DNA in comparison with that of *E. coli* DNA. Labeled *E. coli* DNA was present with the calf DNA during the incubation as an "internal standard". Thus these measurements yield a relatively precise measure of the concentration of complementary sequences in the *E. coli* DNA compared to those in the calf DNA.

The DNA content of the bull sperm is 3.2×10^9 nucleotide pairs per cell [25], and Cairn's measurement [21] gives 4.5×10^6 nucleotide pairs for the size of the *E. coli* genome. The ratio of these numbers is 710, and the ratio of the C_0t for half reaction of the slow part of the calf DNA curve to that for *E. coli* is 690. This establishes with greater accuracy the conclusion drawn from fig. 2 that under these conditions about 60 percent of the calf DNA does not exhibit repeated sequences. If the slowly reassociating sequences were nonrepeating (unique), they should also reassociate accurately since only the precisely complementary strand will be present. In other words, both the hyperchromicity and thermal stability of the reassociated pairs should approach that of native DNA of this fragment size.

Fig. 11 shows the hyperchromic melting curves for very rapidly and very slowly reassociating fractions of calf thymus DNA. The slowly reassociating fraction was prepared by repeated incubation and passage through hydroxyapatite ($C_0t = 100$, 60 °C, 0.12 M phosphate buffer). It was then incubated extensively ($C_0t = 1000$, 60 °C, 0.24 M phosphate buffer), and the reassociated portion was isolated by binding to hydroxyapatite. The C_0t was sufficient only to reassociate about half of the DNA strands. Thus, little concatenation [12] had occurred, and the hyperchromicity had not achieved its final value. Nevertheless, the nonrepetitious fraction has almost 75 percent of the hyperchromicity of native high-molecular-weight DNA and a relatively sharp thermal transition. Much of the difference between this curve and that for native DNA (fig. 7) is due to the fact that the DNA had

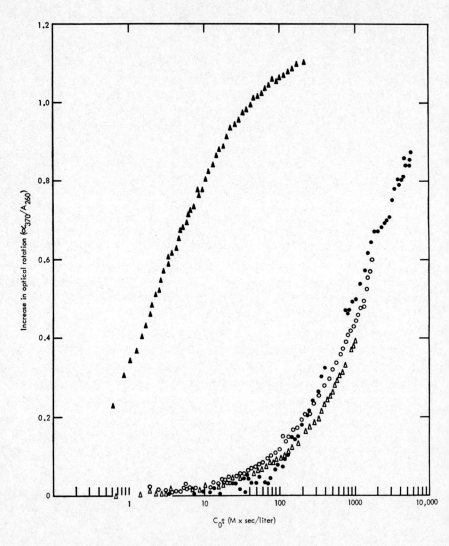

Fig. 10. Measurement by optical rotation at 370 nm (Rudolph recording spectropolarimeter) of the reassociation of *E. coli* DNA and the slowly reassociating fractions of mouse
and calf DNA (0.24 M phosphate buffer, 60°C). Closed triangles, *E. coli* DNA at 0.69
mg/ml in a 10-cm cell. Open triangles, mouse DNA fractionated to remove rapidly reassociating sequences. Open circles, calf thymus DNA fractionated to remove rapidly
reassociating sequences. Closed circles, a second, similar, preparation of calf thymus
DNA. The reaction rates shown here are nearly threefold greater than those shown in
fig. 2 because of the greater salt concentration.

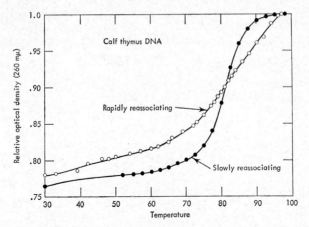

Fig. 11. Hyperchromic melting curves for rapidly reassociating and slowly reassociating fractions of calf thymus DNA in 0.08 M phosphate buffer. Fractions were prepared with hydroxyapatite and reassociated as described in the text.

been sheared before reassociation. Native high-molecular-weight and sheared *E. coli* DNA (fig. 6) differ to about the same extent.

The rapidly reassociating fraction was eluted from the hydroxyapatite in a reassociated state and was diluted to the proper solvent and concentration for the measurement. A very brief incubation was used in this case, and the melting curve is strongly influenced by a particular fraction [32] which reassociates very rapidly, and apparently has a high guanine-cytosine content and melting temperature. A more typical melting curve of a repetitive DNA fraction prepared after a more extensive incubation is shown in fig. 7.

Reassociation of unique sequences of calf DNA was originally observed in our laboratory by means of the spectropolarimeter. This instrument is particularly suitable since a high concentration of DNA is required both for the large $C_0 t$ and for an accurate measurement of the change in optical activity. The slowly reassociating DNA was prepared for this purpose from commercial calf thymus DNA by hydroxyapatite fractionation and lyophilization.

In the region of the spectrum above 300 nm there is a good contrast in specific rotation between the native and denatured states, and little light is adsorbed by the DNA. The reassociation of the purified slow fraction of both mouse and calf thymus DNA occurs at about 1/500 the rate of that of *E. coli* DNA under the same conditions (fig. 10). Within the accuracy of the

measurements this is the expected rate for a 60 percent fraction of mammalian DNA which has no repeated sequences. The reassociation of the unique
sequences of mammalian DNA is therefore confirmed by measurement of
the change in optical rotation.

  Amphiuma is the organism with the largest known genome, having 8×10^{10}
nucleotide pairs of DNA per haploid cell. The expected C_0t for half reassociation in the absence of repetitions would be 80,000 under the conditions
of fig. 2. *Amphiuma* DNA was fractionated on hydroxyapatite after shearing
and incubation ($C_0t = 80$, 50 °C, 0.14 M phosphate buffer). Only 20 percent
of the DNA was recovered in the slowly reassociating fraction. This fraction
was incubated (60 °C, 0.24 M phosphate buffer, 5 mg per milliliter), and
samples were analyzed on hydroxyapatite every few days. It exhibited the
slowest reassociation reaction that has been observed, reaching half reassociation with a C_0t of 20,000. This apparent agreement with the expected
rate cannot be considered definitive until a number of controls are done.
Nevertheless, it appears likely that a significant fraction of the genome of
even this creature is made up of unique sequences.

Patterns of repetition

The rate of reassociation of the DNA of one organism can be evaluated
over the whole course of the reaction (C_0t from 10^{-4} to 10^4). Individual
measurements of reassociation at several concentrations are required, and
fractionation of the DNA is useful. From these measurements the amount
of DNA with various degrees of repetition may be calculated. The result is
a repetition-frequency spectrogram for the DNA of the particular organism,
such as the tentative one for mouse DNA shown in fig. 12. This curve is
correct in its broad aspects but has some indefiniteness in detail. The width
of the peaks results in part from the difficulty of resolving reassociation rates
that differ by less than a factor of 10.

  The large peak at the left on fig. 12 is due to the mouse satellite DNA
[13]. Such a class of DNA molecules which can reassociate with each other
is called a family since the similarity in sequence implies a common origin.
Correspondingly, other classes capable of reassociating are called families
even though the precision of reassociation is less, presumably due to divergence of the members since the formation of the family.

  Repetition-frequency spectra could also be derived for calf DNA and
salmon DNA from figs. 3 and 4. However, they would differ from that of
fig. 12. Neither would show the large isolated peak of 10^6 copies. The calf

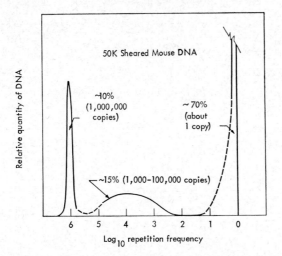

Fig. 12. Spectrogram of the frequency of repetition of nucleotide sequences in the DNA of the mouse. Relative quantity of DNA plotted against the logarithm of the repetition frequency. These data are derived from measurements of the quantity and rate of reassociation of fractions separated on hydroxyapatite. The dashed segments of the curve represent regions of considerable uncertainty.

would show a large broad peak in the region of 10^4 to 10^5 copies (40 percent of the DNA), little if any DNA with a small degree of repetition, and of course a large peak of unique DNA.

What length of DNA sequence has been replicated to form families of repeated sequences? Forty percent of the calf DNA behaves as repeating DNA. The total quantity of repeated sequences is 1.3×10^9 nucleotide pairs per cell. Lengths of DNA totaling 13,000 nucleotide pairs copied 100,000 times would have about the same total quantity and average repetition frequency as the repetitious DNA of the calf. Such a homogeneous set of fragments would have the smallest possible information content that could be present in the repeated DNA of the calf. The situation is known to be more complex, however, and the potential information content of the repetitious DNA fraction is very much greater for the following reasons. (i) The nucleotide sequences of the members of a typical family are similar to each other but not identical. The differences may be of great genetic significance. (ii) A small amount of DNA probably occurs in families made up of long sequences repeated only a few times. (iii) The repeated sequences or fragments of them have been translocated into various parts of the genome, and their location and relationship to their neighbour sequences may be important.

In this regard an observation of Britten and Waring [12] with higher-organism DNA is significant. When 5 to 10 million dalton DNA strands are dissociated and incubated ($C_0t = 1$), large particles form creating a visible haze in solution, and most of the DNA may be sedimented to the bottom of a tube after centrifugation for 5 min at 10,000g. Apparently some regions of most of these DNA fragments are members of families of repeated sequences. Reassociation of the repeated sequences links the fragments into a large network. A number of such measurements with various fragment sizes indicate that repeated sequences are scattered throughout the genome. This extensive interspersion of members of families of repeated sequences may be related to their function. They could, for example, have regulatory or structural roles which would lead to such a distribution. This dispersion may, however, simply represent the degree of translocation of sequence fragments that has occurred during the evolution of higher organisms.

The criteria for repeated sequences

The presence in a DNA solution of a fraction which reassociates rapidly indicates that certain nucleotide sequences are present at a higher concentration than the remainder. If the DNA's were derived from a single organism we can usually conclude that these sequences recur repeatedly in its DNA. The conclusion is essentially certain if the reassociation exhibits the variation of rate with concentration of a second-order collision-controlled reaction.

If such a concentration dependence is not shown other possibilities arise. For example, the rapidly reassociating DNA's could be in closed circular form, the two strands could be cross-linked, or a sequence might occur which contained its own complement and could renature by folding. An example of the latter is the satellite from crab DNA [33] which is principally an alternating sequence of adenine and thymine. In all of these cases, a bimolecular collision is not involved, and the reaction is extremely rapid under optimum conditions for reassociation. With the methods used in this work the rate would be so fast that it could not be observed. Thus, under our usual conditions the observation of a measurable rate of reassociation (faster than that expected from the genome size of the organism) is an almost certain indication of the presence of repeated sequences.

What are the limits of accuracy in the calculation of repetition frequency from a measured rate of DNA reassociation? If only a part of the length of fragments are complementary to each other the rate of reassociation will be reduced. The fraction of the nucleotides which are complementary in typical reassociated pairs of fragments containing repeated sequences is not well known but some limits can be set. The hyperchromicity observed for reassociated repetitive DNA ranges from just less than half up to nearly that characteristic for reassociated bacterial or viral DNA. This evidence implies that half or more of the nucleotides are complementary in typical reassociated repetitive DNA. We believe that under these conditions the reduction in the rate of reassociation is not large. However, the frequency of repetition may be somewhat underestimated in this work.

There is a possibility also of a potential overestimate in the quantity of repeated DNA sequences. When reassociation is assayed with hydroxyapatite, all strands of DNA which contain a sizable double-stranded region will bind. The minimum double-stranded region which will adsorb to hydroxyapatite under the conditions used is not known but is much smaller than the fragment size used. Thus, a certain fraction of the nonrepetitive DNA will be included in the measured repetitive fraction. If the repeated sequences occur in stretches which are long compared with the fragment size, this error will be small. Partly for this reason small fragments are used in this work.

There is evidence that homopolymer clusters occur in DNA [34]. It has not yet been demonstrated that such clusters influence DNA reassociation. The quantity present in the DNA of higher organisms is not known. Nevertheless, in the following paragraphs we attempt an estimate of the maximum effect homopolymer clusters could have on the rate of DNA reassociation.

If the homopolymer clusters were long enough to form a large fraction of the length of a set of DNA fragments they would simply form an extreme example of repeated sequences. The reassociation rate constant would be that shown for the polyuridylic acid and polyadenylic acid pair on fig. 2. The reassociation would appear to be instantaneous under our conditions, except with very low concentrations of DNA.

Very short homopolymer clusters might play two possible roles. They might cause fragments not complementary over the rest of their length to form stable structures which are paired only in the cluster region. These would form a class (not yet observed) of repetitive DNA with very low hypochromicity. Short homopolymer clusters might also increase the rate of nucleation of fragments which were fully complementary and thus increase their rate of reassociation.

The limit for the maximum possible factor of increase is less than twice the number of nucleotides in the homopolymer cluster. This can be seen by making the most favorable assumptions. (i) Nucleations in the homopolymer regions of otherwise noncomplementary fragments do not interfere (that is, they dissociate quickly); (ii) all nucleations occur independently and the rate of reassociation is proportional to the number of "in register" collisions possible; (iii) all possible "registrations" of the homopolymer cluster (twice the number of nucleotides in the cluster) lead to reassociation of major complementary regions if present.

Since condition (i) would not be met for homopolymer clusters longer than 20, the factor of increase in rate must be less than 40. Since condition (iii) is not likely to be met, the factor of increase is probably much less than that. This factor is not large enough to affect the conclusion about the general occurrence of repeated sequences.

Implications of repeated sequences

These studies have revealed new properties of the DNA of higher organisms which must be attributed to the repetition of nucleotide sequences. Some of

TABLE 2

Characteristics of DNA reassociation

Source	
Bacteria	Vertebrates
Viruses	Invertebrates
	Higher plants
	Euglena
	Dinoflagellate
Rate of reassociation	
One rate, inversely proportional to DNA content per cell or particle	Many different rates. Slowest inversely proportional to DNA content per haploid cell. Fastest up to 10^6 times faster
Extent of reassociation	
Excellent, up to 90 percent reformed helices (no strong effect of fragment size)	Good if DNA cut into small fragments. Poor if DNA is of high molecular weight
Stability of reassociated DNA	
Temperature at which strands separate (T_m) almost equal to that of native DNA	Some with T_m near that native DNA and many lower degrees of stability
Particle size of reassociated DNA	
Several times the fragment size due to pairing of free single-stranded ends (concatenation)	Enormous, if DNA fragments are large, due to multiple interconnections (network formation)

these properties are summarized in table 2. In general, more than one-third of the DNA of higher organisms is made up of sequences which recur anywhere from a thousand to a million times per cell. Thus the genetic material is not a collection of different and unrelated genes. A large part is made up of families of sequences in which the similarity must be attributed to common origin.

A minor degree of sequence repetition is to be expected from studies of protein sequences [35]. The hemoglobin group shows similarities in sequences, and these similarities point to common origin of part or all of their structural genes. Trypsin and chymotrypsin also show similarities. There is evidence that, in some cases, different segments of the amino acid sequence of a given protein may have arisen by duplication and insertion of an earlier short segment. In addition to genetic evidence, banding patterns in polytene chromosomes show that gene duplication occurs [36]. The genome sizes of higher organisms range from 10^8 to 10^{11} nucleotide pairs [25]. There is no doubt that a great increase in DNA content has occurred during the evolution of certain species.

These observations suggest that a degree of nucleotide sequence repetition might be observed in the DNA. It must be emphasized that they do not imply that DNA sequence repetition occurs on anything approaching the scale reported here. The very large number of members in the families of repeated sequences remains a most surprising feature for which an explanation must be sought. It may be reasonably predicted that large-scale new patterns of relationship among the proteins await discovery.

Certain minor classes of DNA probably consist of many copies of a short sequence. It appears likely that there are hundreds or thousands of similar ribosomal genes [37] and in certain cells, at least, thousands of similar, if not identical, copies of mitochondrial DNA [38]. Taken together, such classes of DNA do not add up to more than a percent of the total DNA and, compared to the bulk of the repeated sequences, have a relatively low repetition frequency.

If many DNA sequences in a chromosome are similar to each other and adjacent, high rates of unequal crossing-over might occur. Although there is genetic evidence [36] that this occurs, it has not been considered common. Presumably higher organisms are protected from the lethal genetic events [39] that the families of repeated sequences might induce.

There is a certain amount of evidence that repeated sequences are genetically expressed. Pulse-labeled RNA (presumptive messenger) has been hybridized with the DNA of higher organisms [40]. In most of these studies, hybrids were only observed with RNA that was complementary to families of repeated sequences. Due to the small C_0t, hybrids between RNA and nonrepeated DNA sequences of higher organisms apparently did not occur.

The RNA populations made from repeated DNA sequences may have some role (perhaps regulatory) other than as messengers carrying structural information for protein synthesis. However, this is an unlikely (and certainly an unpopular) possibility. A good working hypothesis is that repeated sequences commonly occur in structural genes. In any case, transcription as complementary RNA is direct evidence for the genetic function of at least some of the repeated DNA sequences. In the course of embryonic development and during liver regeneration [40], changes occur in the pattern of types of hybridizable pulse-labeled RNA. These results suggest that during the course of differentiation different families of repeated sequences are expressed at different stages.

Origin and age

The families of repeated sequences range from groups of almost identical copies (for example, mouse satellite DNA) to groups with sufficient diversity that, after reassociation, only structures of low stability are formed among the members. It seems likely that this situation has arisen from large-scale precise duplication of selected sequences, with subsequent divergence caused by mutation and the translocation of segments of certain member sequences. We cannot now describe the history of growth and divergence of any particular family of repeated sequences. However, the few measured properties of the repetitive DNA permit some inferences.

The extensive studies of Hoyer et al. [3, 10, 11] supply a measure of the repeated sequences held in common among different species. Because of the small C_0t used in their work only the reassociation of DNA sequences repeated in each organism was observed. These measurements were carried out at various temperatures [11], and the results were correlated with the period of time after divergence of the lines leading to the modern species [10].

These data show a low average melting temperature if strands of DNA from different species are reassociated. The longer the period after divergence of the species, the greater the reduction in thermal stability. This evidence indicates that the members of families of repeated sequences in the DNA of a species slowly change in nucleotide sequence.

It is an unlikely possibility that all the members of a family of repeated sequences in one organism undergo the same changes. This would involve either very severe selection on all the members or a complex event such as discarding all but one of the members of a family and then multiplying the remaining member 10,000 or 100,000 times. The much more appealing and simpler model is that the nucleotide sequences of the members of the families are not conserved by severe selection. The members may then change slowly and independently of each other leading, after a long period of time, to families with widely divergent members such as are observed.

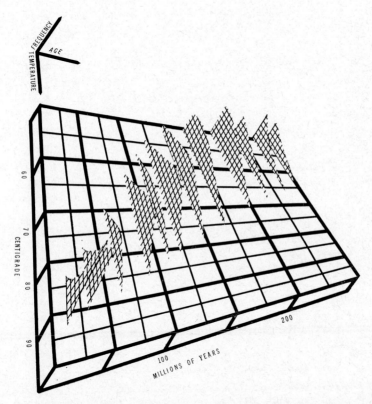

Fig. 13. A schematic diagram intended to suggest the history of the families of repeated sequences now present in the DNA of a modern creature. Each family is supposed to have originated in a sudden event (saltatory replication) at a time in the past shown on the right-hand scale. In the ensuing time, increasing divergence has occurred between the nucleotide sequences of the members of each family of repeated sequences. This divergence is represented on the left-hand scale by the thermal stability of reassociated pairs of DNA strands formed between members of each family. The height of the cross-hatched areas indicates the amount of DNA of a given thermal stability in a family of a particular age. Only a few of a potentially large number of families are indicated. The actual rate of divergence has not yet been well measured.

In addition to the divergence of preexisting families, new families are produced in each species. Analysis of the present data on repetition frequency distribution [14] suggests that they result from relatively sudden events which we have called saltatory replications. Fig. 13 symbolizes the resulting view of the history of families of repeated sequences. Along one axis is the time since formation of the family. Along the other axis is the temperature at which duplexes among the family members dissociate. The third axis represents the number of members. Thus the area of one of the peaks indicates the number of members of the family. The temperature at which the peak occurs is a measure of the extent of sequence difference among the members. The diagram is not intended to be quantitative, although we have used the estimates that are available for vertebrates. The frequency of events and rates of divergence are also probably very different for other phyla.

Even if saltatory replications are as rare as indicated on fig. 13, certain stages of the process may be relatively common. We know nothing of the mechanism of the process, but the following steps seem necessary. (i) A sequence undergoes manyfold replication; (ii) the copies are integrated into the chromosome; (iii) they become associated with a favorable genetic element, and (iv) they are disseminated through the species by natural selection.

Each of the succeeding stages is likely to have a very low probability of occurrence, and thus the actual event of manyfold replication may occur fairly commonly and, in principle, be observable in individual organisms, in analogy to a somatic mutation. It does not seem impossible that, some time in the future, saltations may be artificially introduced into populations as mutations can already be.

Speculation on their function

A concept that is repugnant to us is that about half of the DNA of higher organisms is trivial or permanently inert (on an evolutionary time scale). Furthermore, at least some of the members of DNA families find expression as RNA. We therefore believe that the organization of DNA into families of related sequences will ultimately be found important to the phenotype. However, at present we can only speculate on the actual role of the repeated sequences.

Multiple, nearly exact copies of a gene could provide higher rates of synthesis. This might be true for structural proteins required in large amounts and is very likely true for ribosomal RNA. Multiple similar copies could provide a class of similar protein chains as appear to occur in antibody proteins [41]. However, their role could not be limited to the immune system since they occur in large quantities in the plants and other organisms in which antibodies have not been observed.

The DNA of each vertebrate that has been examined contains some families with 100,000 members or more. This very large number suggests a structural or regulatory role. However, the significance of the very large number might be less direct. It might, for example, raise to a useful level the probability of some rare event such as the translocation of certain DNA sequence fragments into adjacent locations in the genome.

Saltatory replications of genes or gene fragments occurring at infrequent intervals during geologic history might have profound and perhaps delayed results on the course of evolution. In the following quotation Simpson [42] raises some relevant questions with regard to evolutionary history.

The history of life is decidedly nonrandom. This is evident in many features of the record, including such points already discussed as the phenomena of relays and of major replacements at defined times. It is, however, still more striking in two other phenomena copiously documented by fossils. Both have to do with evolutionary trends: first, that the direction of morphological (hence also functional and behavioral) change in a given lineage often continues without significant deviation for long periods of time and, second, that similar or parallel trends often appear either simultaneously or successively in numerous different, usually related, lineages. These phenomena are far from universal; they are not "laws" of evolution; but they are so common and so thoroughly established by concrete evidence that they demand a definite, effective directional force among the evolutionary processes. They rule out any theory of purely random evolution such as the rather naive mutationism that had considerable support earlier in the twentieth century. What directional forces the data do demand, or permit, is one of the most important questions to be asked of the fossil record.

The appearance in a genome of many thousands of copies of a gene could have evolutionary significance. Perhaps not many copies would be actually expressed. Mutation, translocation, and recombination with other genes might yield new genetic potential. If the early effects were selectively advantageous, the repeated DNA sequences could be introduced into the population. The dynamics of selection for this set of genes would be fundamentally altered. Owing to the great multiplicity of copies, their selective elimination might be impossible.

Summary

The rate of reassociation of the complementary strands of DNA of viral and bacterial origin is inversely proportional to the (haploid) DNA content

per cell. However, a large fraction of the DNA of higher organisms reassociates much more rapidly than would be predicted from the DNA content
of each cell. Another fraction appears to reassociate at the expected rate.
It is concluded that certain segments of the DNA are repeated hundreds of
thousands of times. A survey of a number of species indicates that repeated
sequences occur widely and probably universally in the DNA of higher
organisms.

The repeated sequences have been separated from the remaining (unique
sequence) DNA, and their physical properties have been studied. The range
of frequency of repetition is very wide, and there are many degrees of precision of repetition in the DNA of individual organisms. During evolution
the repeated DNA sequences apparently change slowly and thus diverge
from each other. There appears to be some mechanism which, from time
to time, extensively reduplicates certain segments of DNA, replenishing the
redundancy.

Acknowledgements

We thank M. Chamberlin for excellent technical assistance; our colleagues
R. B. Roberts, D. B. Cowie, E. T. Bolton, D. J. Brenner and A. Rake for
direct participation in the preparation of this article; B. H. Hoyer, D. Axelrod and M. Martin for helpful discussion and for several DNA preparations; R. B. Roberts for proposing the schematic diagram (fig. 13); F. R.
Boyd and J. L. England for assistance in the design of the high pressure
pump; and our colleagues and wives for patience and forbearance.

References

[1] J. Marmur, R. Rownd and C. L. Schildkraut, Progr. Nucleic Acid Res. 1 (1963) 231.
[2] B. J. McCarthy and E. T. Bolton, Proc. Natl. Acad. Sci. U.S. 50 (1963) 156.
[3] E. T. Bolton et al., Carnegie Inst. Wash. Year Book 63 (1964) 366; ibid. 62 (1963)
    303; B. H. Hoyer, B. J. McCarthy and E. T. Bolton, Science 144 (1964) 959.
[4] R. C. Warner, J. Biol. Chem. 229 (1957) 711; A. Rich and D. Davies, J. Am. Chem.
    Soc. 78 (1956) 3548.
[5] J. Marmur and D. Lane, Proc. Natl. Acad. Sci. U.S. 46 (1960) 456.
[6] B. D. Hall and S. Spiegelman, ibid. 47 (1961) 137.
[7] E. K. F. Bautz and B. D. Hall, ibid. 48 (1962) 400.
[8] E. T. Bolton and B. J. McCarthy, ibid., p. 1390.
[9] A. P. Nygaard and B. D. Hall, J. Mol. Biol. 9 (1964) 125; D. Gillespie and S. Spiegel-
    man, ibid. 12 (1965) 829.
[10] B. H. Hoyer, E. T. Bolton, B. J. McCarthy and R. B. Roberts, Carnegie Inst. Wash.
    Year Book 63 (1964) 394; in Evolving genes and proteins, V. Bryson and H. Vogel,
    eds. (Academic Press, New York, 1965), p. 581.

[11] M. Martin and B. Hoyer, Biochemistry 5 (1966) 2706; P. M. B. Walker and A. Mc-Laren, J. Mol. Biol. 12 (1965) 394.
[12] R. J. Britten and M. Waring, Carnegie Inst. Wash. Year Book 64 (1965) 316.
[13] M. Waring and R. J. Britten, Science 154 (1966) 791.
[14] R. J. Britten and D. E. Kohne, Carnegie Inst. Wash. Year Book 66 (1967) 73.
[15] R. J. Britten and D. E. Kohne, ibid. 65 (1966) 73.
[16] J. A. Subirana and P. Doty, Biopolymers 4 (1966) 171; J. A. Subirana, ibid., p. 189; J. G. Wetmur and N. Davidson, J. Mol. Biol. 31 (1968) 349; K. J. Thrower and A. R. Peacocke, Biochim. Biophys. Acta 119 (1966) 652.
[17] J. G. Wetmur, thesis, California Institute of Technology, Pasadena (1967).
[18] D. T. Denhardt, Biochem. Biophys. Res. Commun. 23 (1966) 641.
[19] B. J. McCarthy, Bacteriol. Rev. 31 (1967) 215.
[20] G. Bernardi, Nature 206 (1965) 779; P. M. B. Walker and A. McLaren, ibid. 208 (1965) 1175; Y. Miyazawa and C. A. Thomas, Jr., J. Mol. Biol. 11 (1965) 223.
[21] J. Cairns, Cold Spring Harbor Symp. Quant. Biol. 28 (1963) 43.
[22] I. Rubenstein, C. A. Thomas, A. D. Hershey, Proc. Nat. Acad. Sci. U.S. 47 (1961) 1113; J. Cairns, J. Mol. Biol. 3 (1961) 756; E. Burgi and A. D. Hershey, ibid., p. 458.
[23] M. A. Billeter, C. Weissmann, R. C. Warner, J. Mol. Biol. 17 (1966) 145.
[24] P. D. Ross and J. M. Sturtevant, J. Am. Chem. Soc. 84 (1962) 4503.
[25] B. J. McCarthy, in Progr. Nucleic Acid Res. Mol. Biol. 4 (1965) 129; T. Mann, in The biochemistry of semen and of the male reproduction tract (Wiley, New York, 1964), p. 147.
[26] D. E. Kohne and R. Britten, in preparation.
[27] J. Marmur, J. Mol. Biol. 3 (1961) 208; K. I. Berns and C. A. Thomas, Biophys. Soc. Abstr. (1964); B. J. McCarthy and B. H. Hoyer, Proc. Natl. Acad. Sci. U.S. 52 (1964) 914.
[28] B. J. McCarthy and E. T. Bolton, Proc. Natl. Acad. Sci. U.S. 50 (1963) 156; B. H. Hoyer, B. J. McCarthy, E. T. Bolton, Science 144 (1964) 959; E. T. Bolton et al., Carnegie Inst. Wash. Year Book 64 (1965) 314; H. Denis, ibid., p. 455.
[29] B. J. McCarthy and B. H. Hoyer, Proc. Natl. Acad. Sci. U.S. 52 (1964) 915.
[30] E. K. F. Bautz and F. A. Bautz, ibid., p. 1476; T. Kotaka and R. L. Baldwin, J. Mol. Biol. 9 (1964) 323.
[31] A. Rich, Proc. Natl. Acad. Sci. U.S. 46 (1960) 1044; M. N. Lipsett, L. A. Heppel, D. F. Bradley, J. Biol. Chem. 236 (1961) 857; M. N. Lipsett, ibid. 239 (1964) 1256.
[32] E. Polli, G. Corneo, E. Ginelli, P. Bianchi, Biochim. Biophys. Acta 103 (1965) 672.
[33] N. Sueoka, J. Mol. Biol. 3 (1961) 31; ibid. 4 (1962) 161.
[34] H. Kubinski, Z. Opara-Kubinska, W. Szybalski, ibid. 20 (1966) 313.
[35] K. A. Walsh and H. Neurath, Proc. Natl. Acad. Sci. U.S. 52 (1964) 884; K. Brew, T. C. Vanaman, R. L. Hill, J. Biol. Chem. 242 (1967) 3747; E. Freese and A. Yoshida, in Evolving genes and proteins, V. Bryson and H. Vogel, eds. (Academic Press, New York, 1965), p. 341.
[36] C. B. Bridges, J. Heredity 26 (1935) 60.
[37] G. P. Attardi, P. C. Huang, S. Kabat, Proc. Natl. Acad. Sci. U.S. 53 (1965) 1490; ibid. 54 (1965) 185; F. Retossa, K. Atwood, D. Lindsley, S. Spiegelman, Natl. Cancer Inst. Monogr. 23 (1966) 449; S. A. Yankofsky and S. Spiegelman, Proc. Natl. Acad. Sci. U.S. 48 (1962) 1466; H. Wallace and M. Birnstiel, Biochim. Biophys. Acta 114 (1966) 296.
[38] P. Borst, G. Ruttenberg, A. M. Kroon, Biochim. Biophys. Acta 149 (1967) 140, 156; I. B. Dawid and D. R. Wolstenholme, Biophys. J., in press; S. Granick and A. Gibor, in Progr. Nucleic Acid Res. Mol. Biol. 6 (1967) 143.
[39] C. A. Thomas, in Progr. Nucleic Acid Res. Mol. Biol. 5 (1966) 315.
[40] R. B. Church and B. J. McCarthy, J. Mol. Biol. 23 (1967) 459, 477; D. D. Brown

and J. B. Gurdon, ibid. *19* (1966) 399; H. Wallace and M. Birnstiel, Biochim. Biophys. Acta *114* (1966) 296; A. H. Whiteley, B. J. McCarthy, H. R. Whiteley, Proc. Natl. Acad. Sci. U.S. *55* (1966) 519; H. Denis, J. Mol. Biol. *22* (1966) 269, 285; V. R. Glisin, M. V. Glisin, P. Doty, Proc. Natl. Acad. Sci. U.S. *56* (1966) 285; M. Nemer and A. A. Infante, Science *150* (1965) 217; A. Spirin and M. Nemer, ibid., p. 214; M. Crippa, E. Davidson, A. E. Mirsky, Proc. Natl. Acad. Sci. U.S. *57* (1967) 885.

[41] W. Gray, W. Dreyer, L. Hood, Science *155* (1967) 465; S. Cohen and C. Milstein, Nature *214* (1967) 449; G. M. Edelman and J. A. Gally, Proc. Natl. Acad. Sci. U.S. *57* (1967) 356.

[42] G. G. Simpson, This view of life (Harcourt, Brace and World, New York, 1964), p. 164.

TERTIARY STRUCTURE OF RIBONUCLEASE

G. KARTHA, J. BELLO and D. HARKER*

Center for Crystallographic Research, Buffalo, New York, U.S.A.

We started our investigations of the tertiary structure of the enzyme ribo-nuclease in about 1950 at the Protein Structure Project at the Polytechnic Institute of Brooklyn and have continued them in the Biophysics Department of the Roswell Park Memorial Institute since 1959. During the past four years, they have resulted in three-dimensional electron density distributions of the protein molecule in the crystalline state; these maps progressively showed more detail as more X-ray diffraction data at higher resolution were used. Recently we have calculated a map at 2 Å resolution, using the data from the free protein crystal and from seven heavy atom derivative crystals. We believe that this map clearly shows the structure of the molecule. Comparison of this map with the primary structure of the protein, as elucidated by biochemical methods during the past few years, makes it possible to locate the amino-acid residues of which the molecule is made. Many of these residues, including four cystine disulphide bridges, can be located unambiguously. In contrast to myoglobin, the absence of any appreciable amount of α helix makes a complete description of this molecule difficult; it must await a more detailed map. The purpose of this article is to show the general course of the polypeptide chain in the ribonuclease molecule, and to describe some features of biochemical relevance. We have also reason to believe that we have located the active site of this enzyme molecule.

Primary structure

All our X-ray diffraction investigations have been carried out on bovine pancreatic ribonuclease. The primary structure of this protein has been determined [1], so that we know the number, nature and sequence of the amino-acids with a high degree of certainty. This covalent structure is shown in fig. 1 and it can be seen that the protein contains a single polypeptide chain of 124 residues in a specific sequence. This chain is internally cross-linked by four disulphide bridges. The molecule does not possess any free

* Reprinted by permission from Nature *213* (1967) 862.

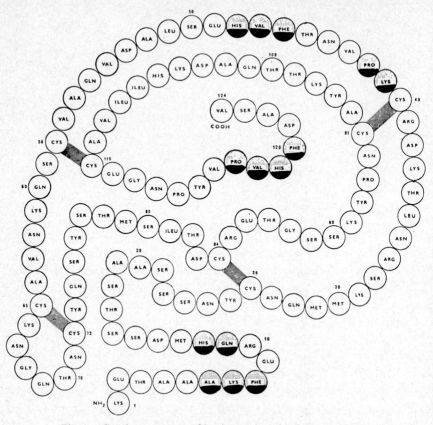

Fig. 1.   Covalent structure of bovine pancreatic ribonuclease [1].

SH or similar groups which facilitate specific heavy atom tagging for X-ray studies. Much information [2] from chemical investigations suggests probable non-covalent interactions between the side groups, when the main chain is folded up into its native configuration; some of these interactions seem essential for the integrity of the active centre.

X-ray data

Crystals of ribonuclease used in these investigations were grown from 55 percent 2-methyl-2,4-pentanediol (MPD) at pH 5.0 (usually in the presence of phosphate buffer) and were monoclinic in space group $P2_1$ with lattice constants shown in table 1. There are two ribonuclease molecules per unit

TABLE 1

Modification II space group $P2_1 \ Z = 2$

		a (Å)	b (Å)	c (Å)	β	Intensity measurements completed 1/19/67 (Å)
Free protein (STD)	$(3P65)$	30.13	38.11	53.29	105.75	1.8
Cis-diglycine pt.	$(9P12)$	29.87	38.39	53.19	105.95	2.0
K_2PtCl_4 + D-serine	$(9P22)$	30.08	38.28	53.23	105.56	2.0
$UO_2(NO_3)_2$ + L-valine	$(9P24)$	30.13	38.18	52.77	105.53	2.4
$UO_2(NO_3)_2$ + arsenazo	$(9P28)$	30.09	38.12	52.81	105.55	2
Sodium arsenate	$(R2-P9)$	30.11	38.01	52.95	105.70	2

cell and the unit cell contains about 40 percent by weight of the solvent. The molecular weight based on the covalent structure is 13,683.

The diffraction data were collected using either the Eulerian cradle [3] on General Electric diffractometers (XRD-3, XRD-6) or, later, the G.E. goniostat. Copper K_α radiation was used ($\lambda = 1.5418$ Å). Earlier, the arcs were manually set from precomputed tables using the stationary crystal, stationary counter method, counting the X-ray quanta at the peak position, with intervening balanced pairs of nickel or cobalt filters (Ross filters); each counting time was 10 sec. Approximate absorption corrections were applied to the intensities. During the past year some of the high resolution data used in these investigations were measured using similar techniques but by setting the arcs by an automated XRD-5 with prepunched cards containing arc settings for reflexions sorted in an order suitable for efficient collection of data. Statistical studies of data collected from the same or similar crystals indicated an accuracy of 4–5 percent in |F| for reflexions within the 3 Å resolution sphere. In the 3–2 Å range, however, the overall reproducibility fell to about 10 per cent, mainly as a result of the comparative weakness of many of the reflexions. A copper target operated at 20 mA and 40 kVp was used as a source of radiation. Wherever possible, measurements were extended to Bijvoet reflexion pairs to take account of anomalous dispersion.

Structure determination

Attempts were made to determine the structure of ribonuclease by means of the method of multiple isomorphous series [4] – a technique which had led to the solution of the myoglobin [5] and lysozyme [6] structures. A search for suitable heavy atom derivatives, conducted over the past few years, revealed a few likely ones, some of which, though satisfactory for low resolution work, were either not easily reproduced or not suitable at 2 Å resolution. Thus not all the derivatives used in computing the 4 Å map except the cis-diglycine platinum (CDG) were used in the later investigations. This (CDG) derivative contained two main heavy atom sites per molecule; anomalous scattering measurements were made in this case. The 2 Å map also included isomorphous derivative data from five other crystals and limited data to 3 Å from tris-(ethylenediamine) platinum (IV) chloride (TEP) for which the anomalous scattering data had been measured earlier.

During the course of the past 3 years, electron density maps were prepared at four resolutions and a summary of these maps is given in table 2. It is emphasized that some of these derivatives had heavy atoms in similar positions, and the similarity of the derivatives in some pairs is so great that one could even have replaced the two independent sets of results by a single one. In fact this was done for the two sets of data for CDG derivatives (soaked in different concentration) in the 2 Å map, even though they were treated as independent derivatives in the 2.4 Å map. Table 3 gives a list of the derivatives, the heavy atom sites and occupancies used in the phase evaluation for the 2 Å map. These heavy atom positions were determined and refined by Patterson, difference Fourier and least squares methods [8–10]. The protein phase angles were evaluated by a combination of iso-morphous series and anomalous dispersion data [11, 12] from the different derivatives, taking care to give proper weights to the phases obtained using any given derivative, and to the reliability of the data from that derivative. The Fourier maps were computed in xy or yz sections and the results printed out in coded form suitable for direct contouring on the output sheets from the IBM 7044 computer.

Many features of the molecule are visible in the 3 Å model computed with 2,340 protein reflexions: (a) the position of the amino end which sticks out quite independently of the rest of the molecule; (b) the depression on the surface of the molecule where the phosphate ion is located (as will be shown later); (c) the position of three S–S bridges of high density near which the main chain density shows the topology expected in these regions. It was also seen that the region near the amino end is partly helical and that the molecules are well separated by regions of very low density.

The 2 Å map involving 7,294 reflexions and data from seven derivatives was computed in sections of constant z at intervals of $x/60$, $y/88$ and $z/112$ – these intervals were so chosen that the printed output had a scale of 1 cm

TABLE 2

Progress in Fourier resolution of electron density maps achieved in this investigation

Date	Reso-lution	No. of "reflexions"* used in map	No. of derivative crystals from which data were collected and used in phase evaluation. At higher resolution only some had their anomalous scattering effects measured	No. for which anomalous scattering was also used	Comments
July 1963	4 Å	1,020	7	7	Showed general shape of molecule and also two regions of high density which looked like S-S bridges
June 1964	3 Å	2,340	In addition to above data from seven derivatives, isomorphous series and anomalous scattering data from CDG and TEP	2	Showed molecular boundary, three S-S bridges, amino end of chain and, later, we could also infer the position of the active site from arsenated RNase
Aug. 1966	2.4 Å	4,895	Data from eight derivative crystals including two separate sets of measurements for CDG	1	All S-S bridges now unambiguous. Whole of main chain could be traced with reasonable certainty, except for a few regions of ambiguity
Dec. 1966	2 Å	7,294	Data for seven crystals; the two separate sets of CDG data were combined into one	1	Ambiguities in tracing the main chain are removed and we can locate all residues with a reasonable degree of certainty

* Each "reflexion" combines information from several derivatives.

TABLE 3

Heavy atom positions and occupancies used in evaluating the 2 Å map

		x	y	z	Occupancy electrons
(1) cis-Diglycine platinum (II) (CDG)					
	1	0.207	0.028	0.936	53.7
	2	0.410	0.500	0.425	54.9
	3	0.285	0.905	0.871	17.9
	4	0.252	0.711	0.008	11.2
	5	0.483	0.676	0.495	7.3
(2) tris-(Ethylenediamine) Pt (IV) (TEP) chloride					
	1	0.126	0.654	0.975	27.4
	2	0.216	0.850	0.990	19.0
(3) D-Serine-Pt (II) complex and Pt(en₃) Cl₄ (PTD)					
Two main sites same as two main	1	0.193	0.036	0.940	39.8
sites of CDG	2	0.416	0.500	0.429	38.1
	3	0.286	0.905	0.875	39.7
	4	0.250	0.715	0.008	11.2
	5	0.467	0.500	0.353	9.0
	6	0.490	0.681	0.495	6.2
(4) Complex of L-valine with uranyl ion					
	1	0.255	0.710	0.008	28.0
	2	0.486	0.682	0.506	21.3
	3	0.253	0.818	0.495	9.5
	4	0.802	0.523	0.061	6.2
	5	0.418	0.495	0.430	9.5
(5) Uranyl complex of 1,8-dihydroxy-2,7-bis (o-arsonaphenylazo)-naphthalene-3,6-disulphonic acid UAZ					
	1	0.264	0.710	0.009	46.5
	2	0.484	0.678	0.504	39.8
	3	0.250	0.808	0.495	19.0
	4	0.797	0.534	0.062	7.3
	5	0.418	0.500	0.430	15.7
	6	0.418	0.558	0.590	11.2
(6) Arsenated RNase NA1					
	1	0.443	0.461	0.388	23.5
(7) Arsenated RNase NA2, basically same as NA1					
	1	0.451	0.462	0.386	24.6

to an angstrom. No $F(000)$ term was added to the series and the contours were drawn on an arbitrary scale of 3 units, with the first contour corresponding to 3. In this scale the disulphide bridges had a density in the range 22–17; the only other peak which had a value of more than 17 in the present map was near the main chain at a region close to the residue 46, and this region had a peak value of 19. The region between the molecules rarely rises to more than contour level of 3 and there are no regions of large negative density.

A schematic drawing of the main chain as deduced from the 2 Å contour map appears in fig. 2. Starting with the amino end, which was quite easily seen even in the 3 Å map, it is possible to proceed along a ribbon of high density corresponding to the main chain by counting along it at appropriate residue distances and using the positions of bulky side groups as a check. In this way it was possible to proceed up to the carboxyl end of the chain, and the four disulphide bridges and their known positions in the amino-acid sequence in fig. 1 acted as a good check in case any adjustments were

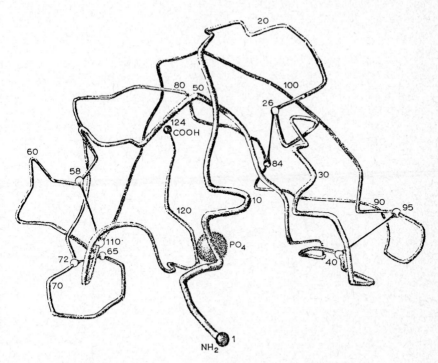

Fig. 2. Schematic diagram of the main chain folding in the ribonuclease molecule (Thanks are due to Mr. John C. Wallace, who drew from the model.)

needed. At the scale (1 cm = 1 Å) to which the map was drawn it was not easy to recognize all the side groups purely from the shape of the contours, but most of the bigger side chains could be identified by their bulk. It is hoped that the larger scale model now under construction will make it possible to recognize independently many of the side groups by their shapes and bulk. The present map has only been used to trace the course of the α-carbon atoms of the main chain assuming the correctness of the covalent structure of fig. 1. No attempt has been made at present to check the correctness of the chemical sequence by identifying the side groups from X-ray data alone.

Conformation of the main chain

The molecule is roughly kidney shaped with approximate dimensions of $38 \times 28 \times 22$ Å with a deep depression in the middle of one side. Two of the disulphide bridges (26–84) and (40–95) are on one side of this construction and the other two (58–110) and (65–72) are close together and on the opposite end. Between these two ends the main chains run in three roughly antiparallel sections; the sections between residues 40–58 and 98–110 running in one direction and region 75–90 running in the opposite direction. Quite clearly the molecule has comparatively low helical content. The only obvious helical segment is about two turns in the region of 5–12 near the amino end and possibly two turns each in the regions 28–35 and 51–58, even though the latter are not quite obvious. There are a couple of other regions where there is a suggestion of helical conformation.

The molecule clearly has a much more exposed structure than myoglobin; the smallest of the dimensions is about 20 Å, and no part of the molecule is shielded from the surrounding medium by more than one layer of main chain. This makes it necessary for some of the hydrophobic side chains which would otherwise be shielded from the solvent to lie near the surface. As a general rule, however, most of the segments which consist predominantly of polar side groups (as, for example, residues 66–71 and 85–91) do show up clearly at the outside of the molecule with their side groups pointing outwards into the solution.

Location of the active site

From an examination of the position of the phosphate ion in crystalline ribonuclease, we now have indirect information about the location of the site of enzymatic activity of the molecule. The position of the phosphate ion was first established by crystallizing ribonuclease from a solution containing, not the usual phosphate group, but the electrostatically very similar arsenate group. The resulting crystal was very closely isomorphous with the phosphate crystal. An electron density difference map computed between the phosphate and arsenate crystals at a resolution of 4 Å, and using the known protein phases, resulted in a single peak per asymmetric unit on a very clear background. The position of the arsenate group was also established independently of any knowledge of the protein phases from a three dimensional difference Patterson map using data from the phosphate and arsenate crystals. The location of this region with respect to the protein molecule showed that the phosphate is embedded in the depression of the kidney-shaped surface of the molecule.

In fact, the isomorphism between the phosphate and the arsenate crystals was so good that, despite the fact that the actual electron difference between an arsenate and phosphate group is not very large, complete three dimensional X-ray data were collected for the arsenate crystals, and these data were used as heavy atom derivative crystal data for protein phase angle evaluation.

Comparison with chemical evidence

Assuming that the active site is indeed near the location of the phosphate ion – an assumption for which there is much chemical evidence and which we can use as a plausible hypothesis – we can locate the regions of the main chain and the amino-acid residues surrounding it. These residues are shaded in fig. 1. These regions occur at different parts of the main chain, and detailed examination of the characteristic arrangement creating the required charge distribution, and other side chain interactions in this region, is likely to throw some light on the nature of the active site. Even though in our map we are dealing with an arrangement which results when a phosphate ion is bound at the active site, it is possible that the mechanism of attachment and the conformation around this region will not be very different when other substrate analogues are bound there.

It is seen that the residues closest to the phosphate are residue 119 near the carboxyl end and residue 12 near the amino end. Both of these are histidines, and much chemical evidence [13, 14] indicates their close relationship with the activity of the molecule [15]. Other residues further out, but which might be of importance, are lysine 7, lysine 41, histidine 48, all of which are reasonably close to the phosphate site. It is also seen from the present map that the amino end residues up to 21 or 22 stand out clearly apart from the rest of the molecule, except for the region near about histidine 12 which comes close to the active site forming its third side. This agrees well with the possibility of cleaving off of this part of the chain by the enzyme subtilisin, leaving the rest of the molecule basically undisturbed.

A fuller description of the ribonuclease molecule and a comparison of the intramolecular relationships of the functional group with chemical evidence for active site groups, intermolecular contacts, and solvent interactions are at present being prepared. Furthermore, we are also preparing a fuller description of the heavy atom "dyes", their preparation and mode of use as well as X-ray diffraction techniques used in the solution of the structure.

We thank the Dean Langmuir Foundation, the Rockefeller Foundation, and the Damon Runyon Foundation for providing the initial support of this project, and the National Science Foundation and the National Institutes of Health for continuing support during recent years. The Roswell Park Memorial Institute and the New York State Department of Health, as well as the Roswell Park Division of Health Research Incorporated, all contributed to the success of this venture by providing space and computing facilities.

In the early days of the work on the structure of ribonuclease, important contributions were made by several scientists no longer connected with this project; among them the following deserve especial mention and thanks: Dr. B. Magdoff, Dr. V. Luzzati, Dr. M. V. King, Dr. A. Tulinsky, Dr. E. von Sydow, Dr. F. H. C. Crick, Dr. T. C. Furnas, jr., Dr. R. Worthington, Dr. A. DeVries, Dr. D. Harris, Dr. H. H. Mills, Dr. R. Parthasarathy and Dr. R. Davis.

We also thank the following assistants: Miss F. Elaine DeJarnette for mounting protein crystals and for collecting most of the X-ray diffraction data during the past six years; Mrs. C. Vincent and Miss K. Go for assisting in data handling and in drawing electron density maps and constructing models; Mrs. Theresa Falzone for excellent assistance in the preparation and for handling the crystals and heavy atom dyes; Misses Elsa Swyers and Sylvia Scapa for preparing some heavy atome dyes; Mrs. Edith Pignataro for collecting data while the project was at the Polytechnic Institute of Brooklyn; Mr. W. G. Weber for constructing, and to some extent designing, the mechanical parts of the single crystal counter X-ray diffractometer.

References

[1] D. G. Smyth, W. H. Stein and S. Moore, J. Biol. Chem., *238* (1963).
[2] C. B. Anfinsen, Brookhaven Symp. Enzyme Models and Enzyme Structure, (1962) 184.
[3] T. C. Furnas and D. Harker, Rev. Sci. Instr., *26* (1955) 449.
[4] D. Harker, Acta Cryst., *9* (1956).
[5] J. C. Kendrew, R. E. Dickerson, B. E. Strandberg, R. G. Hart, D. R. Davies, D. C. Phillips and V. C. Shore, Nature, *185* (1960) 422.
[6] C. C. F. Blake, D. F. Koenig, G. A. Mair, A. C. T. North, D. C. Phillips and V. R. Sarma, Nature, *206* (1965) 757.
[7] G. Kartha, J. Bello, D. Harker and F. E. DeJarnette, Aspects of protein structure, *13* (Academic Press, New York, 1963).
[8] G. Kartha and R. Parthasarathy, Acta Cryst., *18* (1965) 745, 749.
[9] G. Kartha, Acta Cryst., *19* (1965) 883.
[10] B. W. Mathews, Acta Cryst., *20* (1965) 82, 230.
[11] G. Kartha, Am. Cryst. Assoc. Meet., Boseman, July 1964.
[12] A. C. T. North, Acta Cryst., *18* (1965) 212.
[13] E. A. Barnard and W. D. Stein, Biochim. Biophys. Acta *37* (1960) 371.
[14] G. R. Stark, W. H. Stein and S. Moore, J. Biol. Chem., *236* (1961) 436.
[15] F. M. Richards and P. J. Vithayathil, Brookhaven Symp. in Molecular Biology, *13* (1960) 115.

SUBJECT INDEX